MICROCOMPUTERS AND THEIR APPLICATIONS FOR DEVELOPING COUNTRIES

About the Book

Microcomputers are an increasingly important tool in all aspects of development as the need to handle and assimilate vast quantities of information becomes ever more critical for both the international development community and the developing countries. In addition, the microcomputer represents the first significant technological advance that a developing country can assimilate and exploit with a relatively low capital investment and without prior knowledge or involvement in other technologies. Unfortunately this new technology represents not only an opportunity if properly exploited but a threat if ignored. The widespread and increasing incorporation of microcomputers into all aspects of the developed countries represents a major technological advance and an inevitable social change. If a developing country fails to take advantage of the opportunity that microcomputer technology represents, its level of development in relation to developed countries will be significantly lowered.

Organized by the Board on Science and Technology for International Development in response to a request from the U.S. Agency for International Development, this book is an overview of microcomputer applications in developing countries and the issues associated with their use and abuse. The first section of the book is an assessment of the need for microcomputers in development. Written primarily for those in the development field and other computer-literate individuals, the second part is divided into applications in agriculture, health, energy, and municipal management. Policy concerns are addressed in the final section. It discusses the technology transfer that takes place as countries try to establish national computer policies that meet local needs while encouraging creative and useful applications.

Published in cooperation with
the Board on Science and Technology
for International Development,
Office of International Affairs,
National Research Council

MICROCOMPUTERS AND THEIR APPLICATIONS FOR DEVELOPING COUNTRIES

Report of an Ad Hoc Panel on the Use of Microcomputers for Developing Countries

Routledge
Taylor & Francis Group

LONDON AND NEW YORK

First published 1986 by Westview Press, Inc.

Published 2018 by Routledge
52 Vanderbilt Avenue, New York, NY 10017
2 Park Square, Milton Park, Abingdon, Oxon OX14 4RN

Routledge is an imprint of the Taylor & Francis Group, an informa business

Library of Congress Catalog Card Number: 86-50731
ISBN 13: 978-0-367-01080-5 (hbk)
ISBN 13: 978-0-367-16067-8 (pbk)

The Board on Science and Technology for International Development (BOSTID) of the Office of International Affairs addresses a range of issues arising from the ways in which science and technology in developing countries can stimulate and complement the complex processes of social and economic development. It oversees a broad program of activities with scientific organizations in developing countries, and examines ways to apply science and technology to problems of economic and social development through various programs, research grants, studies, advisory committees, workshops, and other mechanisms. BOSTID's Advisory Committee on Technology Innovation publishes topical reviews of technical processes and biological resources of potential importance to developing countries.

This report has been prepared by an ad hoc advisory panel of the Advisory Committee on Technology Innovation, Board on Science and Technology for International Development, Office of International Affairs, National Research Council and has been funded by the Office of the Science Advisor, Agency for International Development, under Grant No. DAN-5538-G-SS-1023-00.

Panel on Microcomputers for Developing Countries

WILLIAM J. LAWLESS, JR., Cognitronics Corporation, Stamford, Connecticut, Chairman

JAMES S. McCULLOUGH, Research Triangle Institute, Research Triangle Park, North Carolina, Vice Chairman

RUTH M. DAVIS, The Pymatuning Group, Inc., Arlington, Virginia

BARBARA DISKIN, International Statistical Program Center, Bureau of the Census, U.S. Department of Commerce, Washington, D.C.

NATHANIEL FIELDS, Institute for International Development, Vienna, Virgina

HARRY HUSKEY, Board of Studies in Computer and Information Sciences, University of California, Santa Cruz, California

DONALD T. LAURIA, Department of Environmental Sciences and Engineering, School of Public Health, University of North Carolina, Chapel Hill, North Carolina

HAROLD LIEBOWITZ, School of Engineering and Applied Sciences, The George Washington University, Washington, D.C.

KURT D. MOSES, International Division, Academy for Educational Development, Washington, D.C.

MOHAN MUNASINGHE, Energy Department, The World Bank, Washington, D.C.

PHILIP F. PALMEDO, International Resources Group, Setauket, New York

KILUBA PEMBAMOTO, McDonald Douglas Corp., Chevy Chase, Maryland

ROBERT TEXTOR, Department of Anthropology, Stanford University, Stanford, California

MICHAEL WEBER, Department of Agricultural Economics, Michigan State University, East Lansing, Michigan

KARL WIIG, Arthur D. Little Inc., Cambridge, Massachusetts

Contributors

T. OWEN CARROLL, W. Averell Harriman College for Policy and Management, State University of New York, Stony Brook, New York

RANEL J. COVERT, Thunder and Associates, Inc., Alexandria, Virginia

RUTH E. DEER, Department of Environmental Sciences and Engineering, School of Public Health, University of North Carolina, Chapel Hill, North Carolina

CHRISTOPHER LAMPTON, Hyattsville, Maryland

GARY L. GARRIOTT, Volunteers in Technical Assistance, Arlington, Virginia

J.A. GUNAWARDENA, University of Peradeniya, Peradeniya, Sri Lanka

STEVEN HARSH, Department of Agricultural Economics, Michigan State University, East Lansing, Michigan

V.K. SAMARANAYAKE, University of Colombo, Colombo, Sri Lanka

National Research Council Staff

MICHAEL DOW, Associate Director, BOSTID
JACK FRITZ, Program Director
F.R. RUSKIN, BOSTID Editor
ELIZABETH McGAFFEY, Program Assistant

Contents

TABLES AND FIGURES xiii
PREFACE xv

PART I
MICROCOMPUTERS AND DEVELOPMENT NEEDS

1 MICROCOMPUTERS: OPPORTUNITIES AND IMPACTS

Information Processing 1
Social Questions 3
Potential Problems 4
Technology Transfer 8
Early Lessons and Research Needs 13

2 GENERIC APPLICATIONS: INFORMATION
AND COMMUNICATION MANAGEMENT

Project Management 15
Information System Concepts 18
Components of a Computer-Based Information System . 20
Communications 29
Conclusion 35

PART II
EXAMPLES OF SECTORAL APPLICATIONS

3 APPLICATIONS IN AGRICULTURE

Introduction 39
Applications Review 42
Selected Examples 49

4 APPLICATIONS IN HEALTH

Introduction 59
Applications Review 61
Selected Examples 85

5 APPLICATIONS IN ENERGY

Introduction 105
Applications Review 105
Selected Examples 132

6   APPLICATIONS IN MUNICIPAL MANAGEMENT

Introduction 155
Applications Review 157

PART III
MAJOR POLICY ISSUES AND THE FUTURE

7   MICROCOMPUTER POLICY DEVELOPMENT:
    A CASE STUDY OF SRI LANKA

The Role of Microcomputers in Development 167
Major Policy Issues 170
A Framework for Computer Policy Analysis 177
Summary 182

8   THE FUTURE

Data Communications 183
Expert Systems 184
Intelligent Interfaces.......................... 186
Video Disc Systems 188

APPENDIX: Microcomputer Hardware and Software ... 191

GLOSSARY 197

REFERENCES 205

INDEX 213

Tables and Figures

TABLES

3-1 Microcomputers for Census and Agricultural
     Research in Developing Countries 45

3-2 Areas Where MSTAT Subprograms Can Be
     Used in Agricultural Research 47

3-3 Applications of Microcomputers in Nepal 53

4-1 Key Parameters of the Various Treatment
     Subprocesses 98

5-1 Energy Planning and Management Models 109

5-2 Examples of Electric Planning Sector Models 113

5-3 Examples of Design Software for Thermal
     Equipment 116

5-4 Examples of Solar Software 117

5-5 Examples of Software Used in Energy
     Conservation Analysis and Planning 120

5-6 Renewable Energy Systems Models 125

5-7 HCR Program Input and Output 131

5-8 Electricity Consumption of Commercial
     Buildings in India 147

FIGURES

2-1 Hierarchical-distributive processing system 21

3-1 Least-cost feed programming algorithm 57

4-1 Organization of the Thai public health system 63

4-2 Countries participating in strengthening
health delivery systems (SHDS) 86

4-3 Study area location 88

4-4 Current health data system 89

4-5 Proposed health data system 91

4-6 Tasks in SRS development 96

4-7 Logic of the branch and bound algorithm
for optimization of wastewater treatment
processes 100

5-1 Typical output for a solar design living
space 118

5-2 Computer output for a flow duration curve 129

5-3 Hierarchical microcomputer modelling
framework for integrated national energy
planning 137

6-1 Property tax records system 159

Preface

In 1984, Robert B. Textor, professor of anthropology at Stanford University, wrote, "We find ourselves in the midst of a revolution, whether we wish to be or not." In these words, Textor describes the information revolution as the third fundamental revolution in human culture-building; the first two were the agricultural and the industrial revolutions. The information revolution, including computers, teleconferencing, remote sensing, satellite communication, fiber-optic communication, robotics, and the emerging computational and video capabilities, may some day be seen by historians as having been as significant in its human impact as these other two great processes that preceded it.

The agricultural revolution took thousands of years to achieve true global impact, and the industrial revolution, almost two centuries. The information revolution, now barely a generation old, is moving with incredible speed. In 1979, the late behavioral and computer scientist Christopher Evans described it in these words:

> Suppose for a moment that the automobile industry had developed at the same rate as computers and over the same period [since World War II]: How much cheaper and more efficient would the current models be?... Today you would be able to buy a Rolls-Royce for $2.75, it would do three million miles to a gallon, and it would deliver enough power to drive the Queen Elizabeth II. And if you were interested in miniaturization, you could place half a dozen of them on a pinhead.

In a recent report to the U.S. Agency for International Development, Lawless and Passman (1985) said:

It is widely recognized that the technological development of microcomputers (i.e., powerful computers based on high speed microprocessors with large internal memories, rapid-access, high-capacity disk storage and with cost a fraction of earlier comparable capacity computers) represents the first significant technological advance which a developing country can assimilate and exploit with a relatively low capital investment and without extensive prior knowledge or involvement in prior technologies....

Unfortunately, this new technology represents not only an opportunity, if properly exploited, but a threat, if ignored. The widespread and increasing incorporation of microcomputers in commerce, industry, and even the social fabric of developed countries represents a major technological advance and an inevitable social change. If a developing country fails to take advantage of the opportunity that microcomputer technology represents, its level of development in relation to developed countries will be significantly lowered.

In 1983, the Bureau for Science and Technology of the U.S. Agency for International Development (AID), asked the Board on Science and Technology for International Development (BOSTID) of the National Research Council to convene a series of symposia to assess the implications of this technology for international development.

Concurrently, in Sri Lanka, the Computer and Information Technology Council (CINTEC) was being formed under the guidance of its chairman, Mohan Munasinghe, to develop a national computer policy and to encourage the use of these technologies. Sri Lanka clearly became an ideal venue for the first symposium in the series. Three areas of specialized application--health, agriculture, and energy--were selected. A proceedings volume, Microcomputers for Development, was published jointly by CINTEC and The National Research Council.

This more comprehensive volume is based on the deliberations of that meeting and other sources. Together with its companion proceedings, it is intended as an introduction to innovative microcomputer applications in developing countries. It is the first of several volumes aimed at practitioners and decision makers involved in planning and implementation of development projects.

The report is organized into three major parts: The Microcomputer and Development Needs, Examples of Sectoral Applications, and Policy Issues and the Future. Each of these sections can be read separately, but together, they will provide the reader with a more complete overview of the field.

As the use of microcomputers spreads, governments may be faced with controlling the influx of microcomputers and software on the one hand, while trying to encourage creative and valuable applications on the other. This process will require an awareness of new technical developments, opportunities, and constraints for microcomputer use in their countries. This publication, therefore, should also be of value to those charged with establishing national computer policy.

William J. Lawless, Jr.
Chairman, Panel on Microcomputers
for Developing Countries

MICROCOMPUTERS AND THEIR APPLICATIONS FOR DEVELOPING COUNTRIES

Microcomputers
and Development Needs

1

Microcomputers: Opportunities and Impacts

INFORMATION PROCESSING

Information is a resource, and like other resources--oil, ore, crops--its value can be enhanced by processing. Traditionally, this has been performed by the human mind, that most efficient and remarkable data-gathering and processing machine. Over the centuries, a class of workers has arisen to perform the task of transforming information from a disorganized form into a structured, useful form. In the hands of an accountant, for instance, fragments of data might be rendered into organized columns of figures; in the mind of a scientist, the same data might coalesce into a theorem. The information produced by this mental processing can, in turn, become data to be processed by others, or it can be used to enhance the processing of other resources. The theorem, for instance, can be used by an engineer to create new machines or structures; the columns of numbers produced by the accountant can enhance an organization's ability to acquire money.

Unlike other resources that are finite in supply and easily exhausted if used imprudently, the amount of available information tends to increase over time, often at a geometric rate. As the reserves of information grow, so does the need for more efficient means of processing it. The sheer quantity of raw data now available in areas such as agriculture, business, energy resources, and government is well beyond the ability of the human mind to process, at least within a reasonable time.

The computer is a machine for processing such information. It may never attain the complexity and sophistication of the human mind, but by performing the unsophisticated, rote activities of an information processing task, such as manipulating columns of numbers, solving mathematical equations, and rearranging text, the computer frees the information worker to make the complex decisions that no computer is yet capable of.

1

For certain applications, the computer, until recently, has been affordable only by those with considerable financial resources. In the developed world, for instance, the computer has become an indispensable tool for government and business. The value of a computer to government or business has traditionally been calculated by comparing the value of the data processed by the computer to the value of the cost of processing by an alternative means. Clearly, if the value imparted to the data by manipulation is greater than the cost of operating the computer, then the extra expense is justifiable.

In the last decade, the cost of owning and operating a computer has decreased significantly. The microcomputer, the product of 25 years of progress in integrated circuit technology, has made it possible for small businesses and research establishments to purchase one or more computers with relatively minor financial investment.

Origin of the Microcomputer

The microcomputer was made possible by the invention, in 1971, at the Intel Corporation of California, of a device called the microprocessor. The microprocessor is an integrated circuit that combines all of the information processing machinery of a computer on a single chip of silicon. In one stroke, the microprocessor allowed the creation of computers that were smaller and cheaper than any that had previously been feasible.

Computer use in the developed world has proceeded at a brisk but haphazard pace for more than a third of a century. In the less developed countries, however, the pace has been somewhat more sluggish. The use of larger computers, mainframes, and minicomputers has been restricted generally to larger corporations and governments. Microcomputers, because of their low cost and portability, allow computers to penetrate sectors where computer use has hitherto been unknown. As the cost of the equipment continues to drop (and it is likely to do so for at least the remainder of this decade), rapid dissemination of microcomputers throughout the developing countries can be expected.

The microcomputer can produce many changes in the developing world. Its advent could change the interaction between developing countries and the international assistance organizations and developed country groups with whom they cooperate. In many areas, it will no longer be necessary for a developing country government to hire outside agencies to perform critical financial analyses. It will be able to do the analysis itself with microcomputers.

Some Initial Cautions

Before discussing microcomputers and their applications, it is well to reflect what microcomputers cannot and should not be expected to do:

- Microcomputers are no substitute for good analysis and thoughtful choices. They may help, however, primarily by simplifying the otherwise tedious examination of a whole range of possibilities or by performing rapid computations when necessary.
- Microcomputers are no replacement for well-trained staff with intelligence and analytic skills. Indeed, the most important feature of the microcomputer may be the power it brings to this group of people to assist them in their duties.
- Microcomputers do not bring sector-specific knowledge to a particular problem, whether in energy, agriculture, or health. "Expert systems" purport to do this, but their widespread practical application is still in the future.

SOCIAL QUESTIONS

Serious questions have been raised about the social implications of introducing microcomputers into developing countries where basic (noncomputerized) communication networks are often weak, large numbers of people may be illiterate, the number of technologically skilled workers is small, unemployment is high, spare parts are often unavailable, electricity is not always reliable or available, foreign exchange may be short, and people often lack more-basic services such as roads, sanitation facilities, and schools. These questions apply to any imported technology and bring up the issue of dependence on industrialized nations and their products.

However, Munasinghe (1984a) points out, "Improving the quality of decision making where management skills are scarce will help to create more jobs at lower levels, rather than making workers redundant. . . . Better application of computers to science and technology will enable the intellectual community to enhance their contribution to national development."

Mainframe computer technology has been in use in many developing countries for years in the operation of airports and banks. Thus many people in the Third World can operate a computer even if they are not skilled in programming. Based on his experience in Egypt, El Kholy confirms that "Many if not most of the developing countries, in which good health and sophisticated technology--aircraft, cars, brain scanners, oil refineries, weather stations, hydroelectric and various industrial plants—exist alongside disease and poverty, have secondary school graduates who can be trained to operate and service a microcomputer" (El Kholy and Mandil, 1983).

In many of these developing countries, mainframe computers at ministries are overburdened. Large amounts of information are collected from field levels, but analysis cannot be scheduled and results are not reported back to the field. The use of microcomputers at intermediate levels, in a district or regional office, for example, helps to solve the problem of processing large amounts of data and the problem of providing timely information. Moreover, by increasing the number of people who use microcomputers, intermediate processing of information helps technology transfer. It trains intermediate-level personnel in planning, logistics, and objective measurement, and allows for a wide diffusion of information among professional personnel.

Once the decision to use a microcomputer has been made, the next step is the choice of software—the programs that it will perform--and hardware--the equipment. It is important to select a system that has software available to carry out the needed tasks; can be operated by the available staff; is capable of performing whatever calculations or functions are required; can run on the local power supply; can operate in local weather conditions (heat, humidity, dust, for example); and has replacement parts and servicing readily available. Further discussion of generic hardware and software is provided in the appendix.

POTENTIAL PROBLEMS

In this section, several practical difficulties that are likely to arise and possible solutions to those problems, or ways in which they can be avoided altogether, are discussed. Chapter 7 includes some of the broader issues that may arise from the introduction of this technology in the Third World and addresses policy questions that should be considered before a framework can be established for computer use in developing countries.

Availability

In most developed countries, purchasing a microcomputer has become relatively simple. In theory, one need only walk into a computer dealership, consult a salesman, and purchase the appropriate system. In practice it is rarely this easy. Nevertheless, computer buyers in North America and Europe take for granted that they will have a wide variety of equipment from which to choose and that they will be able to assemble a system from that equipment without exceptional difficulty or inconvenience.

In developing countries, however, this is not the case. Although certain manufacturers, notably Apple, IBM, and Radio Shack, try to maintain an international presence, established dealerships are rare outside of national capitals; the choice of brands is often limited; and delivery can be time consuming and unreliable, especially when demand for a particular computer exceeds the ability of the dealer or the manufacturer to supply the equipment. Shortages in Third World countries are common. If local purchase is not an option and equipment must be shipped from overseas, customs restrictions and import duties may apply, adding considerably to the overall cost of a system. Some countries limit importation of foreign microcomputers to protect an indigenous computer industry. This ban forces buyers either to purchase locally manufactured equipment, that may not be as up-to-date as that available in other countries, or to seek out black market dealers. Some governments, therefore, are permitting local manufacturers to import systems with the understanding that they will gradually increase the proportion of components manufactured locally.

Service

Most countries do not have an indigenous computer industry; nonetheless, it is desirable to purchase a microcomputer system locally, when possible, if only to guarantee that service will be available when needed. This means, unfortunately, that equipment will often be selected not because it is the best on the market or the most appropriate for the task at hand, but because it is available and can be easily serviced.

Since many of the components in a computer system are delicate, particularly when exposed to environmental extremes, repairs are inevitable. If service is not available locally, equipment may need to be shipped overseas for repairs. Serious consideration must be given to the purchase of a back-up system for use when the primary system is not available. This secondary system should be as close to identical to the primary system as possible, so that it will function with all peripherals and software.

Appropriate Software

Perhaps the most frustrating problem associated with acquiring microcomputers is software compatibility. Microcomputers all require a master control program--the operating system-- usually written in "machine" mathematical language and unique for each type of microprocessor. Only recently some operational systems (CP/M, MS DOS, UNIX) have evolved into de facto standards by being used by the most popular microcomputers.

Furthermore, program development tools such as compilers allow programs written in the popular high-level languages to be used with these available operating systems, so it would appear that there should be no serious compatibility problems. However, commercially offered programs are provided to the user in the machine language format and are therefore not usable on other microprocessors. It also turns out that there are always slight differences in the higher level language used in conjunction with one operating system as opposed to another.

Because the ability to run the desired software is key in the choice of hardware, the dealer who supplies a computer system should also supply software. That software, however, may not always be readily available. Generic software packages, such as word processing, spreadsheets, and database management systems, should be reasonably easy to find, but more esoteric packages, intended for specialized applications, will be more difficult to obtain. To some extent, this is a problem in both developed and developing countries, though it has been lessened in recent years by the increase in numbers of commercial software outlets and mail- order suppliers. Some mail-order software houses in the United States and Europe will send programs to Third World nations, but there is always a risk that programs will be damaged in transit, because of rough handling or exposure to x-rays.

A more serious problem is that suitable software for many Third World applications may not exist. Further, because most of the software industry is concentrated in a few countries, most notably the United States, the bulk of software packages is in English. French and Spanish translations exist for the more popular program manuals, but other language versions are rare. Local development of software could remedy this lack if suitable pack- ages cannot otherwise be obtained. However, software develop- ment is not a trivial undertaking. Sophisticated packages produced by major software publishers are in many instances the result of enormous effort. Whereas less ambitious programs intended to solve specific problems present a less formidable task, the services of a trained programmer are required for all but the most trivial of applications. Translating of existing software into suitable form is easier than creating new packages from scratch but should only be attempted if the source code (that is, the original program in the form written by the programmer) is available. Unfortunately, most software publishers are reluctant to supply source code.

Scarcity of Information

In choosing a microcomputer system, it is important to have up-to-date information about available equipment and software and their operation. In much of the world, this is not a problem. Publication of information about microcomputers has become almost as vital an industry as the production of the computers themselves. Computer users in the developed world take this information resource for granted. In addition, they have access to dozens of microcomputer-oriented magazines, some of them writing about microcomputer systems in general and others specializing in specific brands of machines, such as IBM, Apple, Commodore, and Radio Shack. In some Third World countries, these books and magazines are either unavailable or rare and costly. Further, most are available only in English, French, and Spanish. An alternative in many countries is locally published microcomputer magazines to provide information at low cost in the local language to local purchasers.

Another possible source of inexpensive information is the computer users' group. These groups are informal networks of computer users, often linked by a common interest in a particular brand of equipment or a common end use. Members of these groups are usually willing to share their knowledge about computers and may even offer assistance in the implementation and maintenance of a system. Their newsletters disseminate their knowledge to a wider audience. Contact with users' groups can often be made through local computer dealers or universities.

Training

Despite the emphasis placed by manufacturers on so-called user-friendly systems, hardware/software packages aimed at the nonspecialist user and designed for ease of use, it remains difficult at best to use a microcomputer effectively without some form of training. Often, it may be somewhat challenging to learn a microcomputer application from the manual. However, ideally, the dealer who supplies the computer system will also provide inexpensive training at his facility. If a computer system is installed in a developing country by the local government or an international agency, the project should also allocate funds to train users.

Environmental Problems

Computer equipment is sensitive to environmental extremes, and maintaining conditions conducive to its use can add significantly to the cost of the system. Although microcomputers are less delicate than many larger computers, they are nonetheless subject to overheating and moisture damage. In a country where high humidity and extreme temperatures are common, an effort should be made to minimize the effect of these conditions. Air conditioning may be necessary in some circumstances. The ruggedness of a particular system may even become a primary consideration when making a purchase. Transportable computers, which fold into suitcase-like containers when not in use, are reportedly less liable to environmental damage because they are designed for travel.

Even more important is the availability of a steady and reliable power supply. All computers are subject to damage from sudden power fluctuations. At the very least, a transient power failure can destroy information residing in the computer's Random Access Memory (RAM), and a sudden power spike, a rapid change in voltage, can damage both the computer and certain types of peripheral devices, such as floppy and hard disk drives. Commercial surge protectors are available to provide protection against minor spikes but offer only limited insurance against major power surges. A better, though considerably more expensive, alternative is the Uninterruptible Power Supply (UPS). While this adds considerable cost to a computer system (perhaps even doubling it), it may be preferable to the cost of repairs and equipment replacement and the loss of productivity that would result from frequent system failures. Clearly, the desirability of such a system depends on the state of the local power supply. In an urban setting, the electrical system may be quite reliable; in rural areas, on the other hand, it may not.

If power fluctuations are a major problem, an alternative solution may be the purchase of a lap-sized portable computer. These systems have very low power requirements and will run on batteries if necessary. Already systems are available with the computing power of a 256K IBM PC, and full-sized 80 by 25 character line screens. Eventually, larger battery-powered systems may also become common.

TECHNOLOGY TRANSFER

Local training of microcomputer users provides several benefits. Munasinghe (1984a) writes, "While the transfer of knowledge from abroad is essential for Third World Development, it must not occur at the cost of self-respect and self-reliance." This suggests that when the end users are made aware of the analysis process, they are better able to ensure that the type and amount of data collected is sufficient to answer the questions asked. In addition, if needs change over time, they can modify the software to satisfy current needs. The following examples illustrate how the use of microcomputers has improved the operation of educational, health, and energy-planning institutions.

Environmental Engineering

An example of use in environmental engineering can be found at the Asian Institute of Technology (AIT). The institute offers its computer facilities and software to other users, especially when more advanced and less widely available software is involved. AIT was established to help meet the need for advanced engineering education in Asia; it is supported by more than 30 Asian and western countries and draws its students from approximately 20 countries in the Asian and Pacific region. In addition to an IBM 3031 mainframe computer, AIT has several microcomputers in its various units. Many programs written for the mainframe have been or are being rewritten for the micro by students and faculty members from various countries.

The Division of Environmental Engineering (DEE), for example, offers a program stressing computer applications at the doctoral level; it also offers short courses and seminars on specific subjects. Emphasis in these courses has recently been switched to the use of microcomputers from the use of mainframes. The DEE uses microcomputers in teaching and research; it offers courses in microcomputer use, software development, and specific use of programs in advanced engineering design.

From its experience with software for the planning and design of water-related systems, the DEE has developed a number of principles for software development for practicing engineers: It must be convenient to use and amenable to modification by the user; data input must be easy; and data requirements must be as low as possible to achieve a useful output. A major benefit of design programs is their ability to compare alternatives both from the standpoint of design and cost.

Many prepackaged programs are not precisely compatible with the conditions in a given developing country. Software should, therefore, allow the program to be modified to reflect local conditions and priorities. This requirement is often met by preparing the program in modular format; this allows modification or replacement of individual program modules without the need to rewrite software outside the module being replaced. Recognizing that users may need to modify software, AIT includes training in rewriting packaged programs in its courses.

Health Services

Family Health International (FHI) in Research Triangle Park, North Carolina provides microcomputers and training in their use to developing countries. It has carried out technology transfer programs in Indonesia, Thailand, and Tunisia for population and demographic survey work. Before this program, information had been collected in these countries and sent to mainframe computers in other countries for analysis. Thus, developing country workers were cut off from the results of their data collection, opportunities for verification of data were lost, and results were slow to return to the originators of the data.

In contrast, the use of microcomputers has had a number of positive results. First, research and evaluation skills were upgraded significantly. Second, local personnel became skilled in programming, data cleaning, editing, and analysis. Third, local data processing and analysis capacity shortened the time lapse between data collection and the generation of information needed by health care providers and managers and allowed new questions to be explored more easily. And finally, a totally integrated system of hardware, software, documentation, installation, and training was constructed.

Once a regional center has been recognized as one in which both background and research goals would benefit from the acquisition of a computer, an on-site evaluation of the proposed site is conducted. If appropriate, hardware, supplies, technical manuals, and supplementary literature are ordered. The hardware is "hot-staged" at FHI by being assembled and operated under destination-country electrical conditions for a minimum of six weeks. During this hot-staging phase, plans for in-country hardware maintenance and training are completed.

When a machine has arrived at the proposed site, a staff member installs the system and conducts an intensive training session. This training is directed toward diverse developing world users with little or no computer experience. The purpose of the course is to prepare local personnel for all aspects of the operation and use of the microcomputer. Sessions are tailored to each recipient center, based on skill and experience levels of key personnel. Topics offered include fundamental computing concepts; sound data management practices and techniques; operating procedures; procedures for site administration, including system maintenance and care and control of supplies; correct and efficient use of application programs; exception processing and error recovery; and, where applicable, an introduction to the BASIC programming language.

The integrated nature of data collection and analysis with automated data processing is also stressed. Thus, generally six months after the machine has been installed, advanced training in research methodology, biostatistics, and epidemiology, and their relation to the microcomputer is also provided. System use can be reviewed at the same time and recommendations made on future training efforts. Future training could include, for example, development of customized programs in BASIC or FORTRAN, a more detailed look at the operating system and system procedure language, and how best to develop and analyze survey data systematically using a statistical package.

The microcomputers placed to date have tended to encourage and increase the level of research activity in the centers for which they have been provided. Clearly, the development of a strong data-processing capacity is necessary for an independent research center; therefore microcomputers are an important consideration in long-range planning for these institutions.

Energy Planning

Modelling and simulation are the basic tools of planning and systems analysis. The earliest national energy planning efforts in developing countries were primarily assessments; they included no plans for action. Data and information were processed in the United States on mainframe systems. Subsequent assistance was directed toward enhancing planning skills among developing country staff through training programs and seminars. However, no attempt was made to ensure that development of planning skills was accompanied by corresponding microcomputer applied analysis skills. Only recently have development assistance projects included specific placement of microcomputers in developing countries and the training of staff to do analysis on the machines.

There are several examples of AID-supported national planning that illustrate microcomputer-based applications. Energy/Development International (E/DI) carried out a novel National Investment Planning Project for the Comision Nacional de Politica Energetica in the Dominican Republic. This work sought to compare energy supply projects and conservation efforts relating macroeconomic and energy investment considerations to energy prices; thus it transcended general national energy planning, which encompasses only broad national energy strategies.

Unexpectedly, the preliminary results of the initial phase supported coal importation and final-fuel use projects over development of indigenous energy resources such as hydroelectric power. A more general set of conclusions indicate that national staff could participate effectively in the planning and analysis process.

In another example, Development Sciences Inc. (DSI) developed a software package to enhance the planning capabilities of the Ministry of Energy and Mines (MEM) in Morocco (Gordon et al.,

1984). The system links energy supply, demand, and other baseline conditions to the national energy system and planners' expectations about the future. The model then determined the performance of the portfolio over a 20-year period in terms of energy demand and supply balances and its financial and manpower resource requirements. With this information, the planner was then able to return to the beginning of the planning cycle, redefine the portfolio, change project characteristics, or modify alternatives such as prices or imports and note their impact.

Seminars and workshops are becoming popular mechanisms for providing training on the use of microcomputers for energy planning. ACRES American organized workshops in national energy planning and management in East and West Africa in early 1984. In this instance, the microcomputer was utilized directly to carry out energy supply-demand balances, demand projection, and other typical planning activities. A modified version of the World Bank assessment approach formed the basis of instruction, with worksheets using LOTUS 1-2-3 spreadsheet software. A hypothetical country, Terrania, provided the database for energy balances and impact assessment of programs in industrial energy conservation and other practical exercises.

Microcomputers have also been used to determine the national economic impact of energy investments. For example, in the last few years, energy sector investment in Colombia has increased from 6 percent of total investment to almost 15 percent of investment, and the resulting debt service threatens the balance-of-payments position of the country. As part of a national energy study for Colombia, sponsored in part by the Inter-American Development Bank, a preliminary analysis of balance of payments was performed using a modified version of a World Bank model (Carroll, 1983). This national accounts forecasting model was implemented with LOTUS 1-2-3 on the IBM-PC. Other microcomputer based models for macroeconomic forecasting and energy demand estimation have been prepared by MetaSystems and applied in World Bank studies for Bangladesh and other countries.

As part of an industrial productivity project in Egypt, Georgia Institute of Technology plans to use microcomputers to develop an information and data system. Data collected will represent all aspects of industrial production in Egypt, including employment, sales, financing, and energy consumption patterns. Volunteers in Technical Assistance (VITA) is also providing a renewable energy documentation system available as a turnkey operation. Here the microcomputer is utilized for cataloging and record search, and some 8,000 documents listed in the catalog are available on microfiche. VITA also offers a series of workshops during the year to train developing country individuals in all aspects of information and data system design and implementation.

EARLY LESSONS AND RESEARCH NEEDS

Information system concepts and the need to consider all links in the chain are perhaps even more important in developing country settings than in industrial nations. This is especially true in regard to supporting databases and the end user's analytical ability. Without these, there can still be a rapid infusion of foreign techology and equipment; however, there can be no long-term institutionalization nor fundamental contribution to problem solving. Furthermore, overinvestment in imported microcomputer hardware and software has a high opportunity cost if it competes with investments in developing and maintaining databases and training informed decision makers.

Each of the four data processing approaches typically available in developing countries (manual, mainframe computer, microcomputer, and programmable calculator) has clear advantages and disadvantages. Because microcomputers and programmable calculators are relatively new technologies, technicians are still learning how best to use them. Manual tabulation is probably still the best choice for small data processing tasks done only once or infrequently. Programmable calculators, when available, are best used for mathematical calculations performed in a field office by a single technician. The microcomputer is probably best adapted to a central or regional office to do tasks that require repetitive processing of moderate amounts of data, and the mainframe is useful for very large data processing tasks where turnaround time is not critical (Weber et al., 1983).

While microcomputers are proving to be relatively robust for a developing country setting (and they are improving continually), there are still important technological weaknesses, including intolerance to the power supply fluctuations, very high humidity, dusty environments, and static-electricity-prone locations previously mentioned. Good in-country service and repair facilities are not appearing as rapidly as sales outlets. Although this is probably a transitory problem, it may require several years to overcome.

Finally, it should be noted that it is necessary to develop software for actual problems of developing countries at the same time that commercial software is utilized to the maximum. This has two implications: information sharing and evaluation of existing software is very important for developing country users; and specialized software development and adaption are also necessary.

The contrast between models and software in the industrialized countries and realities in developing countries suggests some research areas:

- Increased productivity and economic independence is an assumed goal of national economic development. How can the introduction of microcomputers be measured in terms of the productivity of the various end use sectors, for example? Is it enough to assume that the increased efficiency and data processing capabilities of government ministries and government-sponsored projects will have a beneficial effect on the economy as a whole? What can be said about more private sector participation? Further research is required to analyze how the benefits of the applications of the microcomputer as they are currently defined can trickle down to the farm and village level.

- In the industrialized countries, there is widespread use of the microcomputer in all sectors. Part of this results from previous patterns of dissemination of technology in the industrialized economies. Few rural residents in developing countries are even aware of the potential of microcomputers, and usage is concentrated in government bureaus and experimental fields. What are the implications of this unbalanced characteristic in the spread of the use of microcomputers in developing countries? For industrialized nations, it has been argued that the spread of microcomputers has led to a relative decentralization of power. Can one make the same argument for developing countries?

- What is the nature of the demand for microcomputer technology in developing countries? What are the factors that will favor the long-term institutionalization of microcomputer technology? If microcomputers are misapplied, their benefits can be quickly subverted by a long list of political, economic, and social factors.

It is important to emphasize that growing experience with the technology will enable developing country users to decide on the appropriate responses to these concerns. Many of the problems faced in trying to bring microcomputer hardware to developing countries will be alleviated as the technology matures. By stressing cooperation in software development, developing countries can minimize problems caused by the inappropriate use of software designed for industrial nation applications. Finally, as the cost of the technology continues to decrease the spread of microcomputers even to rural and village areas will become routine, rather than a hard financial decision.

2

Generic Applications:
Information and Communication
Management

PROJECT MANAGEMENT

According to the Development Projects Management Center of the University of Maryland, "Management can be viewed as a human process in which resources are mobilized and productively combined to accomplish meaningful results under conditions of partial control and a constantly changing environment." Management is an adaptive science; the strategies of management must be flexible enough to function in a changing political, social, and technological environment. "It involves a continuous cycle of planning and replanning in which managers and technical staff work together to track accomplishments against planned results, respond to unexpected changes, and incorporate lessons learned from experience." Throughout the process of adaptive management, there is a constant need to collect, analyze, and act on information. It is an information-intensive undertaking, one in which data takes on a very high relative value in terms of its significance as an input. (Ingle and Smith, 1983).

Good management practices are often the key to successful operation of institutions and projects. In developing countries, management can become quite complex, requiring a coordination of efforts on a number of levels in an environment of scarce resources: lack of communications infrastructure, lack of data collection or processing techniques, and lack of funds and trained personnel. The real effect of these constraints has been a disappointing performance of management objectives in a number of cases. The World Bank's World Development Report 1983 takes as its central theme the need to promote more efficient management practices in operating ministries of low-income countries.

Major deficiencies exist in the management of development efforts. During planning and design of projects, while emphasis is frequently given to economic and technical considerations,

15

implementation requirements are often underestimated or, sometimes, totally ignored. As a result, the management and technical staff responsible for implementing and replanning are frequently hard-pressed to keep the effort on schedule and within budget.

For seven years or so, microcomputers have played a role in improving and sustaining the management performance of institutions and development projects. The Development Projects Management Center's "Microcomputers and Agricultural Management Survey" for example, documented that microcomputers were being used in more than 50 AID-funded projects in more than 30 countries by 1982. Applications ranging from simple text editing to complex financial control can strengthen previously inadequate management practices. Microcomputers can provide executives, technicians, and administrative staff quick access to information, thereby saving time in handling many routine office functions and assisting with complicated—and heretofore not feasible—appraisals and analyses.

Advantages of using microcomputers include:

- Timeliness. Improved turnaround time for data and reports is possible. Budget reports, for example, can be made earlier than was possible before microcomputers were used, increasing their value as input for ongoing planning.
- Improved Data Collection and Analysis. The availability of database software can make data organization and analysis more timely and effective. Microcomputers are beneficial here in two respects. First they aid in the construction of relevant databases by simplifying the process of data collection and storage. Second, microcomputer software eliminates a good deal of human error associated with data input and computation.
- Highlighting Important Information. Reports generated using microcomputers allow managers to focus on only the most important information. Information that is voluminous and difficult to decipher can be broken down and clarified. Microcomputers can lower the cost of obtaining truly relevant information and therefore act as an incentive for managers to use more information as a basis for their decision making.
- Improved Presentation of Cost Implications. By visually displaying budgets and estimates of costs, microcomputers can have a direct beneficial effect on the planning process.
- Bureaucratic Impact. All bureaucracies are concerned with their long-term survival and the limits of their responsibilities. There is an incentive to master the tools

that microcomputers offer, since other organizations are doing the same. Competition is raised to a new level, away from political or personal antagonism and towards a more constructive level where information is at the center of the debate.

● Other Social Effects. Microcomputers may broaden opportunity for some groups. For example, the case study of the PROCALFER agricultural project (discussed in the next chapter) reveals that microcomputers can have beneficial secondary effects. In that setting, it was observed that women professionals received new status as a result of their willingness to interact with the new technology. Because men had little experience with typing, for a variety of traditional reasons, women were at an advantage in learning microcomputer operation and were the first to discover the power of the microcomputer.

There are three prerequisites for choosing microcomputers for a project:

● All project participants involved, both those from the developing country and outside consultants, must agree on the objectives of the project.

● The strategies chosen to reach the objectives must be realistic, taking into account the conditions under which the project will be carried out and the resources available. All participants must agree to work according to the strategies chosen.

● Each person involved in the work must be aware of individual responsibilities within the overall plan and must be willing to carry them out in the appropriate time frame.

The microcomputer can enhance the organizational abilities of those involved in the project. It can be used to collect objectives submitted by various parties, to cross-index them, and to prepare an agreed-upon list of final objectives. Possible strategies under discussion can be stored in the microcomputer. When the appropriate strategies have been selected, a microcomputer can be used to prepare a matrix for the overall time frame. Individual areas of responsibility can be put into a flowchart for each sector of the project team. Based on these flowcharts, realistic estimates of personnel and resource needs can be prepared on the microcomputer. Finally, the responsibilities of each individual can be identified, and computer-generated charts can be used to secure understanding and agreement from concerned policymakers and participants before the project begins.

Once the project is underway, the microcomputer can continue to make a contribution. Managers must monitor schedules, and if necessary, make revisions easily. Each sector of the flowchart can be further subdivided to indicate staff, equipment, and resource needs in a particular time segment. Individuals with appropriate skills can be aware, within narrow time limits, of when their contributions will be expected to take place. Because information stored in the microcomputer can be continuously updated to reflect changes in scheduling, specialists can be notified immediately of changes that will alter the timing of their contribution to the project. Equipment and supplies can be ordered to arrive as needed.

The microcomputer can be used to store actual expenditures and actual time used in carrying out various aspects of the project. Continuous interim reports can be prepared, permitting updated forecasts and changes to the basic workplan. Inequities between forecast and actual time and expenditures can then be recognized early enough to compensate for them.

Finally, when the project nears completion, provision must be made for assessment of the work, so that appropriate changes in strategy can be made in future projects to help project managers improve their planning capacity and evaluate alternative strategies.

INFORMATION SYSTEM CONCEPTS

The various types of decision makers in developing countries seeking to obtain information include: biological, technical, and social science researchers; extension agents; teachers; marketing and business firms; managers; farmers; administrators; and planners and policymakers. The types of decisions being made may be quite different, but all of these decision makers are attempting to bring about a more productive and socially beneficial economic system. Their decisions are affected by the quality and timeliness of supporting information.

In general, the value of information is the value of the change in decision behavior caused by the information less the cost of information (Davis, 1974). In other words, given a set of possible decisions, a decision maker will act on the basis of information at hand. If new information causes a different decision to be made, the value of the new information is the difference in value between the outcome of the old decision and that of the new decision less the cost of obtaining the information.

To support decision making, information systems must be carefully designed. A comprehensive information system has four main components: (1) descriptive information, (2) diagnostic information, (3) predictive information, and (4) prescriptive information.

Descriptive information describes "what is." An accounting system in its initial stage provides descriptive information. It can indicate profitability, return on investment, and other financial factors. Other sources of descriptive information include commodity price reports, stock and crop records, weather forecasts, and energy resources.

Diagnostic information describes "what is wrong." It reflects a fact-value conflict. What might be the cause for a low rate of return on an investment when compared with other similar operations? Why was the price received lower than average? Why does variety x perform so poorly under minor drought stress? This diagnostic information helps identify strengths and weaknesses and suggests what should be changed.

Predictive information describes "what if" situations. This is the process of looking to the future. It explores the impact of alternative strategies. Careful generation of predictive information will greatly enhance the likelihood of success for a business person, researcher, or other type of decision maker.

Prescriptive information describes "what should be done." It identifies a plan for the future that the decision makers will try to implement to improve their overall operation. It is the process of making and carrying out decisions and is based on the supply of adequate information through the entire decision making process.

Of course, the values in information systems are determined by the goals of the decision maker. For example, in collecting diagnostic information, what is "good" or "bad" is partly influenced by the values and expectations of the end user. Likewise, in predictive information, the alternatives to be considered are also influenced by the goals of the decision maker. The goals of a business reflect the basic values of the manager. As an example, a hospital may have as its major goal serving the community, making a profit, or training new doctors. The chosen goal influences the information needed. It is also influenced by other factors (how "good" is my product; how much credit can I obtain) facing the manager.

From the preceding discussion, it is clear how important is defining what information is needed for decision making. Once determined, the appropriate processing system and supporting data can be quantified. The opposite approach--collecting data with no clear indication of what information is needed--will only result in wasted resources. First, it becomes necessary to define the components of an information system.

COMPONENTS OF A COMPUTER-BASED INFORMATION SYSTEM

The highest priority should be given to software development, sharing, and evaluation. If software is to be developed for application in less developed countries, what should be its characteristics? In answering this question, we need to consider the five basic components of a computerized information system:

- Hardware
- Software
- Supporting databases
- The end user's analytical ability
- The sales, service, and training support systems.

Hardware

Because of the rapid advances in computer hardware, this technology is now within affordable reach of a significant proportion of the world's population. Stephen Knight of Bell Laboratories recently stated that by the year 2000 he expects to have a memory chip capable of storing 40 million bytes. Other projections are even more optimistic. For data transmission, the technological advances look equally promising.

In many industrialized countries, there is a trend towards a hierarchical distributive processing system, which uses different sizes and classes of computers to handle different tasks (see Figure 2-1). The largest computers in this system, (mainframe computers) are used to maintain very large databases or those databases that must be shared by the entire user community, or both. They are also used to perform large and complex computational tasks.

Minicomputers, the next level in the system, are used to store databases needed by a subset of the user community. They are also used extensively for computational purposes. Microcomputers, the lowest level of the system, are used to store the local databases and supply computational power needed for decision making and problem solving at that level. A distributive system places processing capabilities and data storage and retrieval functions at the appropriate level so that decision makers have access to the needed information on a cost-effective basis.

A study of the potential role of microcomputers in developing country statistical offices (Diskin et al., 1983) concluded that they can probably best be used in combination with other computing equipment, in order to allow access to additional resources.

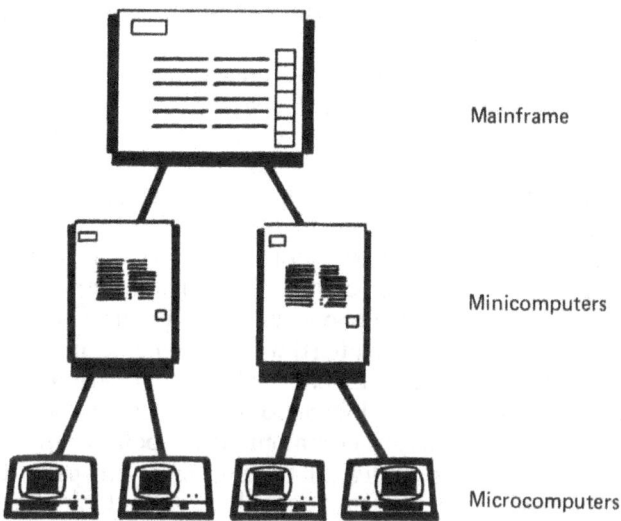

Mainframe

Minicomputers

Microcomputers

Figure 2-1   Hierarchical-distributive processing system

These might include things like faster line printers or the greater storage capacity commonly found on mainframe computers and minicomputers. This observation is based on response from the user (National Statistical Offices) survey which showed that for many applications it is desirable to be able to transfer data from one microcomputer to another or to a mainframe computer for further processing. This allows the microcomputer to be dedicated to a specific task, such as data entry or analysis, instead of forcing it to be a general-purpose machine.

However, it must be pointed out that microcomputer technology is advancing so rapidly that the storage capacity, speed, and capabilities of today's microcomputer far exceed that of microcomputers as recently as two and three years ago.

In storage technology, the video disc is a new development that appears to have a major economic advantage over current storage techniques. A double-sided video disc can hold nearly 20 billion bits of data, 100 times more data than the current magnetic media. Furthermore, it can be mass reproduced. The chief disadvantage of the video disc is that once the data has been recorded, it remains on the disk; it is not possible to erase it and reuse the disk. This may change in the near future and only poses a disadvantage if the data changes rapidly. For storage of historical data (for example, weather data, disease patterns), the advantages of the video disc are obvious.

Overall, these technological advances mean that the computers at all levels of a hierarchical distributive processing system will become more powerful and less costly. It is entirely possible that within a few years the smallest computers in the system will have more capacity than the scientific machines currently found on university campuses.

For this system to function, there must be a reliable, rapid, efficient, and cost-effective communications system. With the recent deregulation of the telephone industry in the United States, the world of electronic communications has become more competitive, and potential for improvements in communication systems is promising. There are strong indications that communication (including voice) between major communication points will be done digitally. This movement to digital transmission will permit better utilization of new transmission technologies, including satellites and optical fiber.

Satellite transmission technology is also advancing rapidly. It is now possible to acquire a small (1 meter) downlink disk for a few hundred dollars. Emerging competition in the satellite communications area should lower the cost of transmission significantly. By time-sharing a transmission signal, a researcher could receive massive amounts of data which are unique to his particular topic in only a fraction of a second. The cost of acquiring this data would be only a small part of the current cost of receiving similar information through standard telephone transmission means or the postal service. More will be said about communication in a subsequent section.

Software

Software is a major cost component; indeed it is likely that software costs will exceed hardware costs for many applications in the years ahead. As noted, there have been major improvements in word processing, electronic worksheet, and database management programs. Sector specific software is described in subsequent chapters. Practical uses for generic software are described below.

Database Management

Of the generic types of software, perhaps none has a wider range of applications than the database management system. It is at once the simplest of software applications and the most powerful, the most clearly defined in function and yet the most flexible. Because most database management systems are, to some extent, programmable by the user, they may be tailored to individual needs. There are few sectors that could not find some value in their application.

Simply put, a database management system accepts information typed on the keyboard of a microcomputer and stores that information on magnetic media, usually either a diskette or hard disk, depending on the quantity of information that needs to be stored. (A hard disk, having greater storage capacity than a diskette, would be appropriate for very large files of information, containing many hundreds of thousands of characters.) After the information is stored, the database management software will, on request, sort it into a desired order, such as alphabetical, numerical, or chronological, and retrieve it either in part or in its entirety. Because the information is stored in compact electronic form, it takes up little storage space. And because it is in a form that is readable by computer, it may be retrieved rapidly and selectively.

The following example will show how database management software can be applied to the needs of a typical industrialized country user.

A small business can acquire a microcomputer with a 15-megabyte hard disk drive, letter-quality printer, and database management software to store, on the hard disk, a list of customers' names and addresses for mass mailing of advertising material. Storing the list of names is a relatively straightforward typing task. Some software packages place a limit, typically in the neighborhood of 65,000 items, on the number of items that can be placed in a single list.

It is only after the names and addresses are stored, however, that the true potential of database management becomes apparent. Under control of the database management software, the list may be retrieved at any time from the hard disk and reviewed on the video display of the computer; alternatively, it may be printed on the system printer as often as desired. Before it is printed or reviewed, the list may be sorted into any of several possible sequences: alphabetically according to name or city, for instance, or chronologically according to the date of the customer's most recent purchase, if that information has been included in the list. (If our hypothetical business were based in the United States, the list might be ordered numerically according to zip code, thus imposing a logical geographical arrangement on the names.) Further, only a portion of the list may be retrieved at any time; subsidiary lists may be printed containing only those customers residing in a given city, or who have last names beginning with a particular letter, or who have made purchases since a particular month, or any combination. From time to time, new names can be added to the list and sorted into current sequence. The names of customers who have not made purchases in a certain number of months can be periodically deleted.

Doubtless, our hypothetical proprietor will discover in time that the database software will store other lists as well: tax records, merchandise in inventory, employee records, and so forth. And a truly sophisticated database package will do more than simply store and retrieve this information; it will add columns of figures, compare and contrast related lists, and print out formatted reports based on the data in those lists. If the database software is part of an integrated software system that also includes a spread-sheet, it may be possible to move information directly from the storage file to the spreadsheet matrix, with no need for retyping the data. If a word processor is included in the package, the names and addresses from our hypothetical mailing list might be used to create form letters.

Most businesses, organizations, and agencies clearly can find some use for computerized database management. For instance, a hospital could use such a system to store the names and addresses of patients, with separate files maintained for each patient's medical records. If a rare blood type were needed for an emergency transfusion, a list of patients with that type could be printed and consulted for possible donors. A school might store the names and grades of students on the computer and calculate grade point averages with the database software.

Of course, not all forms of data lend themselves to this sort of computerized storage and retrieval. The computer will not impose order on an intrinsically chaotic system. An office that has up-dated its records haphazardly before purchasing a microcomputer will, in many instances, continue to update its records haphazardly (although some observers argue that the very presence of a computer in an office provides a strong psychological motivation toward organization).

Spreadsheets

Data storage and organization are not the only reasons for using a computer; even those who deal with small amounts of information can make effective use of a microcomputer system. For instance, decison-making processes that might normally be approached in an intuitive, haphazard manner can be given structure through programs such as electronic spreadsheets. These spreadsheets can simulate the dynamic relationships between real world events in such a way that the consequences of hypothetical actions can be studied in a risk-free environment. Budgets, in particular, are amenable to spreadsheet modelling. A projected budget can be placed in a spreadsheet file, and relationships between items in the budget can be defined; then selected items can be altered to gauge the effect of such changes on the budget as whole. Various expenditures can be increased or decreased; factors affecting income can be manipulated. For each alteration, the spreadsheet software will update all dependent factors and immediately display the result. The best combination of factors, and the most effective budget, can be determined.

The computer, of course, cannot make decisions, but using spreadsheets may save hours or even weeks of time that would otherwise be spent computing budget items by hand. The spreadsheet is a significant tool for anyone who must make decisions based on theoretical projections--businessmen, administrators, scientists, engineers, and planners, for example.

Word Processing

Like database management and spreadsheet software, word processing is useful in a variety of business situations and environments, though it is somewhat more specific in purpose and less flexible in its applications. Word processing software is used to produce documents. That alone may be sufficient to justify the purchase of a microcomputer by many organizations.

With word processing software, documents may be created and fully revised before a copy is ever committed to paper. Editing may be performed on the video display of the computer, and documents can be stored on disk as long as needed. Multiple copies can be printed, and small changes can easily be made from one copy to the next to add a personalized touch to mass-produced documents such as form letters. Because of the decreased need for retyping, document development time is considerably reduced, flawless copies can be produced, and revisions are simple and effortless.

Communications

A fourth type of software program is communications, or terminal, software. Such programs are commonly used in conjunction with a modem, a peripheral device that is used to transmit electronic data, usually via telephone lines. Communications software mediates the transmission of data, words, numbers, even programs, between computers, often over long distances.

Communications programs can link computers to databases containing information on a variety of subjects. (These databases are discussed in the next section.) Electronic mail, the transmission of messages between computer users or between a computer user and a service that will forward hard copy of computer-transmitted information to recipients in distant cities, is also gaining in popularity. Electronic mail may be particularly applicable to the needs of developing countries, where surface mail may be unsatisfactory and international mail excruciatingly slow.

The types of software mentioned thus far in this chapter--database management, spreadsheet, word processing, and communications—have broad applications in any number of fields, and all should be of value to users in less developed countries. Specific uses for computers in such sectors as agriculture, energy planning, public management, and health care will be discussed in greater detail in the following chapters.

Supporting Databases

It has been said that a new on-line database is created each day, but most of these are in the developed industrialized countries, particularly the United States. The Cuadra Associates' Directory of On Line Databases lists over 1,800 databases available through 270 vendors. Access to these databases is difficult, and at best uneven outside the developed countries, with the additional irritant for researchers from developing countries who may have to search databases in the United States to get information about their own country.

This has led to the current debate on trans-border data flows. Data obtained by expatriate researchers may be taken out of the country to be analyzed at computer facilities at home universities. As a result, many countries heavily tax or prohibit access to international retrieval and data-processing services. A 1983 report of the Rome-based International Bureau of Informatics (IBI) concluded that "many countries have yet to formulate data-flow policies." In addition, the report found that the primary users of international database access were manufacturing and trading companies, air and sea transportation firms, and financial information services. Databases established for these users are not usually easily available to in-country users.

There are two general kinds of databases: (1) reference databases, which provide bibliographic information on external documents relevant to a given search; and (2) source databases, which provide the specific information desired, usually without additional searches. In reference databases, the extra step of locating and soliciting the desired document requires additional time and money. A case in point is DEVELOP, a database containing more than 15,000 pieces of information on appropriate technologies, development agency project experiences, specialized personnel, and new techniques in health, education, agriculture, water and sanitation, and business and industry. A U.S. company, Control Data Corporation of Minneapolis, began constructing this database four years ago, giving grants to about a dozen development organizations. But despite a "mail-in" search service and arrangement in which on-line information seekers can barter their new information for existing DEVELOP information, it is little used outside the United States because the cost and inadequacies of telecommunications systems in other parts of the world make access prohibitive. The need to contact a computerized database in the United States also creates a situation of dependency; some African development information officials have termed this "development information domination" (Murphy, 1984).

The supporting database in most countries (developing as well as developed) for agricultural, health, and energy applications is inadequate. Many of the problems faced by researchers and planners require data not generally available. This includes internal operational data as well as external data, such as energy resources and weather forecasts. For much of the needed biological and technical research to increase agricultural productivity, for example, databases do not exist because the on-station and on-farm experimentation necessary to generate this data has not been done. In many other cases the existing databases are personalized by individual researchers and too frequently do not become part of the public record. Finally, there is always a financial constraint in starting and maintaining adequate databases.

The capability exists for building decision-support models and information systems that are far beyond the decision makers' ability to supply the necessary data to support these models. However, most of the problems of inadequate databases can be solved not by the application of new computer technology but by increasing the availability of well-trained people who can conceptualize, establish, and maintain statistical reporting services.

The End User's Analytical Ability

If microcomputer-based information systems are to be successfully used, the end user's (decision maker's) analytical skills may need improvement. Several developing country institutions, U.S. and European universities, international research centers, and other organizations are conducting workshops to train end users on the fundamentals of the computers. However, the impact of these workshops has yet to be felt.

To use either the general or special-purpose software effectively, the user must have adequate conceptual skills to understand how and why to apply specific software to specific problems. For example, in working with an economic problem, the user needs to know whether capital budgeting, cash flow planning, linear programming, or some other analysis technique is appropriate for the situation at hand. Because it is also necessary to understand how to use each technique, a major educational effort is required to provide potential users with the required skills.

A number of individuals and institutions recognize the need for training in conceptual and analytic skills. A group at Texas A&M University is developing and deploying a microcomputer budget management system that will have potential applications in agricultural research, extension, and teaching activities in developing as well as developed countries (McGrann et al., 1984). Recognizing the need for underlying analytical skills, the software development team has written a supporting educational manual to help users increase their knowledge of enterprise budgeting and whole-farm economic and financial analysis. At Michigan State University the team developing MSTAT (a microcomputer package to design, manage, and analyze agricultural experiments) has written a brief statistical guide to assist scientists in learning (or relearning) experimental design and statistical analysis techniques in MSTAT. One agricultural scientist has observed that in Bolivia, agricultural researchers themselves began to appreciate their analytical weaknesses when using microcomputer hardware and software to analyze experiments.

> The first reaction is one of surprise at completing so much work in such a short time. There is then a bit of dismay at the large amount of paper generated. This is followed by a rather detailed interpretation of the data. At this point, there is usually a large demand for statistical counseling. Although the results are in standard form, we find that most technicians have never understood statistics. With the introduction of a new method of calculation, they feel free to express their doubts. Researchers who have been doing their own calculations for years now ask questions like "What does the F value mean?" or "If it's significant, what do I do?" (Stillwell et al., 1983)

Finally, it should be noted that while microcomputers can be fun to work with, they are not likely to change the basic motivation of users. If a researcher or manager currently has a strong dislike for record keeping and detailed analysis of problems, it is unlikely the computer will eliminate this feeling. The end user's expectations of the computer must go beyond the belief that it will free him of all paper work and hard analytical thinking.

Sales, Service, and Training Support System

To make effective use of a computer system, decision makers need an adequate technology and software support system to assist them. This system in developing countries is changing and improving rapidly in almost all countries, particularly in Asia. Yet continued efforts are also necessary to provide the readily available and effective repair, training, and sales services requred by widespread use of microcomputer equipment.

Disken et al. (1983) concluded that the two biggest problems for developing countries' statistical offices in the use of microcomputers are power supply and maintenance. They also observe that considerable frustration and idle microcomputer systems have resulted from hardware and software problems and inadequate training and support. Another important insight from their study is that many problems in the past using mainframe computers in national statistical offices were related to the sales, service, and training support systems. Their list of problem areas, in priority order, was:

1. Access to the computer;
2. Hardware failures followed by delays in repair;
3. Rapid personnel turnover with subsequent need for training new personnel;
4. Management of the computer center;
5. Cost involved in equipment rental and maintenance;
6. Electrical failures;
7. Printer speed;
8. Reliance on cards for program development;
9. Insufficient number of disk packs;
10. Scarcity of sector-specific packaged software; and
11. Lack of technical documentation in the country language.

COMMUNICATIONS

What is required to establish a computer-mediated communication? On the user end, a computer or terminal with appropriate communication software is attached to a modem (modulator-demodulator), a device that converts digital signals leaving the computer or terminal into a modulated (analog) audio waveform that travels through the telephone system to another computer (the "host"), which converts the audio signals back to digital signals.

The host contains the software and message-storage capacity and is the "hub" of the electronic mail system. If the computers and terminals are not physically close to each other (for example, in the same city), then establishing the interconnection is complex. The telephone system itself may consist of a combination of wires, microwave relay links, computers, and satellite communication links. This is always the case when communication is international, and is often true when geographically dispersed within a single country. INTELSAT, the International Telecommunications Satellite Organization, handles almost all international voice, data, telex, and video communication.

While it is technically feasible to make a long distance call via INTELSAT to the computer itself, in practice it is much cheaper and easier to establish communication first with a "value-added network," which automatically connects the desired computer on command. Value-added networks use "packet switching," a digital-processing technique that allows many digital "conversations" to share the same channel while providing very high-quality communication, thanks to built-in error-checking routines. Each "packet" is a burst of digital data addressed to a specific computer. If not acknowledged, the packet is discarded and the transmitting computer is instructed to repeat the transmission. Value-added networks may be thought of as pipelines into which many separate data transmissions are combined at one end and emerge at the other end correctly differentiated.

Three separate methods of using data communications to perform database searches or to send and retrieve electronic mail exist: long-distance telephone, local telephone, and telex links. If the user is located in a developing country and the database or electronic mail service is in the United States, the user must establish a communication link with a U.S.-based computer.

Each of the three methods is illustrated here with an example from Volunteers in Technical Assistance (VITA), an international voluntary technical assistance organization. VITA is using microcomputer technology extensively for both communication and information management internally (within its headquarters offices near Washington, D.C.) and in field offices around the world. Because VITA needs to locate many volunteer experts, often scattered around the United States and Canada, extensive use is made of electronic mail service within North America as well as with selected field offices.

One such electronic mail service is the Electronic Information Exchange Service (EIES), which consists of advanced communication software resident on a computer at the New Jersey Institute of Technology in Newark, New Jersey. Through value-added networks like GTE's Telenet, EIES is connected to hundreds of users throughout the world. A sister development organization in the

United States, Partnership for Productivity, is actively nurturing a user group on EIES for trade and investment purposes called "CARINET" (the original members were located in Caribbean-area countries). EIES contains all the software and command syntax to send and retrieve messages, as well as to perform more advanced features like writing and editing common documents among two or more users (potentially useful for composing joint proposals, for example). But its primary function is as a message service. Messages addressed to a particular person or group are sent to the EIES computer where they are stored until the addressee computer or terminal signs on and retrieves them.

Long Distance Telephone Link

The first type of communication involves making a long-distance telephone call to the United States, connecting the value-added network Telenet to the user's terminal or computer. Readily available computer software will dial this number direct if long-distance direct lines exist; otherwise, the user must contact the international operator at the national exchange to place the call. The user's terminal or computer responds with its own tone while communication between the two is being established. Once communication has been established, the user will be allowed to receive waiting messages (including confirmations that previous messages sent have been received by the respective parties). New messages can then be typed in (or loaded from computer memory or media storage such as diskettes) and sent as desired.

Local Telephone Link

The second method is identical to the first except that a local telephone call is made because an in-country Telenet node exists. This is possible if the country's national telecommunications regulatory or operational body has paid Telenet (or an independent contractor) to establish packet-generating hardware or software or both between that nation's international satellite service and domestic service. Cost for installation is reportedly less than U.S. $100,000. Telenet nodes now exist in more than 50 countries. In Asia these include Hong Kong, Japan, Korea, the Philippines, Singapore, Taiwan, and Thailand. VITA has made frequent use of the Telenet node in Thailand to maintain close contact with its field operation there. In this case, the users at both ends only have to make local telephone calls, thus saving on long-distance charges, although local monthly access costs are still substantial. The accumulated experience to date, however, indicates on-line access using a value-added network local node connection is generally about three times cheaper than using regular long distance channels.

Telex Link

The third possibility is use of the international telex network to gain entry to the value-added data network. Telex, widespread throughout the world, uses a different encoding standard ("Baudot") than data does ("ASCII") for character generation, and generally does not require conditioned lines for reliable service (unlike much data communication). However, it is quite slow (about six times slower than the minimum standard data transmission rate of 300 baud or 30 characters per second).

Costs and Other Constraints

For the first and the second methods above, costs include long-distance or local telephone charges, or both, on each end and Telenet time access charges for each user. In addition, EIES charges its customers a flat-rate monthly charge, regardless of on-line time used. For the telex method, standard telex charges are applicable.

If, instead of an electronic mail service, a database vendor (such as DIALOG, which sells access to nearly 200 databases) is contacted, costs are still appreciable and may take many forms. These costs include subscription fees, connect time fees, printing and storage fees, and others. Therefore, a complete accounting of all fees, is necessary before such a link can be economically established.

It should be emphasized that unless a given country has a local Telenet or Tymnet node, through which the user automatically obtains computer access without any operator intervention required, prior instruction of the local operator as well as with the national exchange is required. In addition, computer modems made in the United States for U.S. equipment will usually not work when connected to a value-added network node outside of North America. The fundamental problem is that the United States, Canada, and Mexico operate on a different standard (the "Bell" standard) from the rest of the world, which uses standards set by the International Telegraph and Telephone Consultative Committee (CCITT), a UN-sponsored committee based in Geneva.

A New Model for Communication Via Satellite

Some of the problems and issues impeding widespread use of international data communications have been discussed. Mating INTELSAT (and domestic satellite systems) with national (often dated) telephone systems to allow wider use can doubtless be accomplished eventually, given enough political will and financial resources. But there is another model worth consideration that could provide low-cost, reliable international data communications for small users who cannot afford access to the larger system.

The communications satellite is the key element in international data transfer. Both for international and domestic service, geostationary satellites (orbiting approximately 22,500 miles above the earth, they appear motionless to an observer on the earth) can cover approximately one-third of the earth at a time. Given the distance and frequencies used, the cost and complexity of full-function earth stations is very high (USAID's Rural Satellite Program, for example, recently delivered an INTELSAT earth station to rural Peru for a reported $200,000 installed cost). Yet for many development applications, the requirement is merely to send messages back and forth, not to enjoy access to the full range of voice, data, telex, video, and facsimile services the geostationary satellites make available.

One can therefore conceive of a satellite in low-earth polar orbit that passes over every point on the earth two to four times a day and that acts as a kind of "electronic mailbox in the sky," collecting ("uploading") messages as it passes above ground stations in its path, storing them in its on-board computer memory, and then transmitting ("downloading") them to the correct addressee when that particular ground station comes into view of the satellite.

There are a number of advantages such a system offers for development-related communications, which tend to be international in nature:

- Ground stations can be relatively inexpensive (US $2,000-3,000 each) because the satellite is only a few hundred miles in orbit, as opposed to many thousand miles, and because the choice of operating frequencies (very high frequency--VHF, or ultra high frequency--UHF) does not require the use of parabolic "dish" antennas, but only simple vertical "whips." Because the system is portable, it is useful for applications such as disaster relief and monitoring supply logistics for refugee camps. Stations can run on battery or solar power, so independence from the vagaries of electrical power is possible.

- The system does not depend on the existing telecommunications infrastructure with its attendant technical and institutional problems, though clearly it would require regulation in its use for communicating within a country to and from rural areas and towns and villages that will never have reliable international data communication capability.

- It uses the same sophisticated "packet switching" techniques described earlier, making high-quality, error-free communication possible.

- It allows for automatic operation, so that the user need not be present to upload or download messages.

- It provides low-cost communication capability for those groups and organizations that cannot now regularly use the international system.

- While the satellite's on-board memory would be too small to make resident any substantial database for on-line searching (in addition to the limited 10-15 minute time "window" per pass), messages describing the boundaries of such searches could easily be transmitted to another party on the system with access to those databases who could more easily carry out the search. Given current costs, use of an intermediary organization for this purpose would be advisable even in the present situation.

- South-south communication would be facilitated, lessening dependence on the industrialized nations.

Such a "flying mailbox" dubbed "PACSAT," is now being designed by VITA volunteers and another U.S. voluntary nonprofit group, the Radio Amateur Satellite Corporation (AMSAT), which has a 25-year history (counting its predecessors) of building and launching low-cost satellites that ride "piggyback" as secondary payloads on U.S. and European launch vehicles.

Low-earth-orbiting (LEO) systems were suggested in 1982 by Yash Pal, Secretary-General of UNISPACE '82 in his "Proposal for an Orbital Postman to Meet Some of the Communications Needs of the United Nations System." Earlier, at a 1981 workshop on computer-based conferencing systems for developing countries sponsored by the International Development Research Centre in Ottawa, Canada, discussion centered around LEO satellites for management purposes. The discussion concluded that development of a LEO system should be well within the capabilities of a group of developing countries, or even individual countries; it further suggested that "a common interest group, concerned with information on technology for development, be identified."

In 1983, VITA and AMSAT joined forces to develop the PACSAT system, with a targeted launch date of late 1986 or early

1987 through the "Get Away Special" program sponsored by NASA on the space shuttle. Teams have already successfully designed and built an experimental version of PACSAT that was launched aboard a British scientific satellite on March 1, 1984.

Various organizations were encouraged to become ground station recipients during the proof-of-concept phase to be undertaken during 1987, and more than a hundred have responded. While PACSAT (and LEO systems in general) will never replace INTELSAT, it could complement INTELSAT where its applicability is most difficult: reliable, low-cost service to small users effectively isolated from the existing international communications infrastructure.

CONCLUSION

As Lauffer (1984) has pointed out, the role of communications, and particularly, data communications, is clear:

...Communication has become a vital need for collective entities and communities. Societies as a whole cannot survive today if they are not properly informed. Self-reliance, cultural identity, freedom, independence, respect for human dignity, mutual aid, participation in the reshaping of the environment, these are some of the non-material aspirations we all seek through communication. But higher productivity, better crops, enhanced efficiency and competition, improved health, appropriate marketing conditions, proper use of irrigation facilities are also objectives, among many others, which cannot be achieved without adequate communication and the provision of needed data.

PART II

Examples of
Sectoral Applications

3

Applications in Agriculture

INTRODUCTION

The purpose of this chapter is to chronicle current uses of microcomputers in agriculture in developing countries as well as more advanced uses in the United States--irrigation control, for example. The hope is that it will encourage microcomputer use in this sector and bring about applications that have not previously been considered.

Computer specialists are well aware of the needs of the agricultural sector. For example, the University of Florida's Institute of Food and Agricultural Sciences has produced an inventory of more than 1,700 programs (many of them for programmable calculators rather than microcomputers) available through extension services in the United States (Strain and Simmons, 1984). Most of these programs are intended for specific purposes: analyses of tobacco irrigation costs and returns, citrus grove record-keeping systems, and calculation of sawmill profit margin. There are literally hundreds of programs of this sort in the inventory, geared to production of various crops, irrigation needs of specific climates, and financial requirements of local tax laws. A very few would seem, on the basis of the brief descriptions included in the inventory, to be applicable to the needs of most farmers in developing countries--for example, farm business record-keeping systems and calculations of loan-cost repayment schedules. Most of these programs are written in either BASIC or FORTRAN, computer programming languages that are portable to a wide variety of microcomputers; thus they are not tied to a specific computer brand.

The use of microcomputers in agriculture is limited because the needs of farmers vary widely from place to place. The computer needs of farmers in tropical developing countries are not necessarily the same as those of farmers in temperate

39

industrialized countries; the crops, the soils in which these crops are grown, and the climate are all different. Thus, most of the software applicable to the needs of farmers in the United States, which includes most of the programs listed in the University of Florida inventory, would need to be rewritten to apply to the needs of farmers in developing countries.

Computer Use In U.S. Agriculture

When digital computers first appeared on U.S. university campuses for general use in the 1950s, agricultural scientists quickly realized their potential, and developed software to analyze agricultural problems. Some of the original statistical and operations research packages developed by agricultural scientists, such as the SAS statistical package, are still widely used today.

The use of computers to address real-world problems of farmers was soon to follow. The daily production records (DHIA) and farm accounting systems (for example, TELFARM and ELFAC) were among the first successes in this area. These systems used batch-operated computers, and employed the mail service for delivering information to and from the data processing center. Many of these projects are still operating. Although improvements have been made in the format of the records, the method for processing data remains basically unchanged.

The use of batch-operated mainframe computers in teaching programs was largely directed toward simulation models. Several simulation games were developed to teach students economic and business management concepts. Most of the games developed were either farm or agri-business models, but a few (such as animal genetics) were simulation models that related to technical areas of agriculture.

As time-share computers became more common in the mid-1960s, Michigan State (the TELPLAN Project) and other universities developed software and a delivery system for farm and extension office use. Examples of problems that the software addresses include ration balancing, financial management, and farm planning. The audience reached by these systems still remains relatively small in proportion to the total farming community.

Several modules covering concepts such as supply and demand and basic production economics were developed and used as a supplement to classroom lectures and assigned readings. At the University of Illinois, the PLATO system was used to help teach genetics. With the genetics model, a student could "breed" several generations of fruit flies in a few hours. To conduct these same breeding experiments with actual fruit flies would require several weeks.

Although its potential was great, Computer Aided Instruction (CAI) also had some associated problems. First, it is very time-consuming to develop educational materials for CAI, and faculty members were often unwilling to invest the time needed to develop a good CAI module. Also, running a CAI module on a time-share computer was somewhat expensive. Many departments did not have the financial resources to make extensive use of CAI to enhance the teaching program.

A product of large-scale integration (LSI) research efforts was the development of programmable calculators in the late 1970s. Suddenly, farmers had available to them a low-cost, portable, and personalized computer capacity. Many universities, realizing the potential of these calculators, developed numerous models for these devices. Acceptance by the farmer has been encouraging. However, this technology has limitations; in particular, the size of the unit limits the magnitude of the problem that can be solved. Furthermore, its limited data storage capacity precludes it from being used in many applications, such as maintaining a record system.

Currently, there is movement for using microcomputers in teaching programs. Most of the colleges of agriculture at major universities have microcomputer teaching laboratories used to teach the basic concepts of computer technology and to run applications programs. The applications programs are mainly decision aids that are similar to the ones developed earlier for time-share computers and programmable calculators. Looking to the future, it is envisioned that microcomputers will be used increasingly in a CAI environment, since the microcomputer has greatly changed the relative economics of using this approach.

The newer microcomputer systems have also made possible the complex modelling of agricultural systems. Pest management models, agricultural sector models, plant physiology models, and machinery selection and design models are just a few of the agricultural system models that have been developed. These have given the agricultural scientist a greater understanding of how the varying agricultural systems function and how these interact with the other agricultural and nonagricultural systems.

The microcomputer has also changed the methods by which research data are collected and analyzed. With the "lap computer" and/or data loggers, scientists are collecting data directly from field experiments in digital form, which is then fed to larger microcomputer systems and quickly analyzed with statistical packages such as MSTAT.

APPLICATIONS REVIEW

Agricultural Management

Reduced initial investment cost, greater accessibility, and relative ease of operation have made microcomputer ownership a popular goal of many agricultural researchers and administrators in developing countries. Starting some 5-7 years ago, when microcomputers were just beginning to appear on a broad scale in industrial countries, many developing country agriculturalists were introduced to this technology by traveling consultants and longterm technical assistance counterparts who owned their own computers. Some developing country scholars studying in the industrial countries acquired firsthand knowledge and sometimes ownership of this new technology. As hardware and software distributorships began to appear in large cities of many developing countries, interest grew rapidly.

In 1982, two workshops were held in the United States to review the use of microcomputers in developing countries. Agricultural research applications were the focus of an international conference held at Michigan State University. Management applications were the focus of a workshop held in Washington, D.C., organized by the Development Project Management Center of the Office of International Cooperation and Development, United States Department of Agriculture. These two workshops produced reports that were widely distributed (Weber et al., 1983, and Development Project Management Center, 1982). These reports provided interested users in developing countries with new ideas and insights about the existing and potential uses of microcomputers. Documented cases of Third World research and management applications were reported, with a view to giving readers ideas for new and different applications.

When these workshops were held, there were virtually no reported uses of microcomputers in developing countries for agricultural extension and teaching purposes. This has changed somewhat in the past two years, with microcomputers being used both at the ministry level and selectively in the field, for the most part by consultants.

Since 1982, more powerful agricultural research and analysis software programs have been developed and are starting up in developing countries. FARMAP, a package for facilitating the collection and analysis of farm management data, has been developed by FAO; another FAO program has been developed for agricultural project analysis (Cappi and Giffen, 1982). MULBUD, a multienterprise, multiyear budgeting program, has been developed at the Australian National University in collaboration with the International Council for Research in Agroforestry (ICRAF) and the International Development Research Centre (IDRC) (Etherington and Matthews, 1984). MSTAT, a program to design, manage, and analyze agricultural experiments, has been developed by scientists from the Agricultural University of Norway and Michigan State University (Freed and Weber, 1984). Researchers at the Oklahoma Climatological Survey have been involved in a program to develop software and place microcomputer-based systems for climatological data processing in Third World countries (Eddy et al., 1984).

Microcomputers are also being used in financial, budget management, project management, and monitoring areas. For example, the Planning Division of the Ministry of Agriculture in Tunisia has used electronic spreadsheets for forward planning of food requirements (Christie, 1984). The Agricultural Ministry in Kenya has made rather extensive use of microcomputers to improve budgeting and financial management (Pinckney et al., 1982). This experience, however, has had some problems. The staff of the Harvard Institute for International Development, working with Kenyan counterparts, have set the following criteria for successful use of microcomputer technology, and have made substantial progress only on items (1) and (2).

This project has been successful in beginning to institutionalize the use of microcomputers for financial management. Institutionalization will be complete when all of the following have taken place: (1) Kenyans themselves demand the output from the machines; (2) all operations are done by Kenyans; (3) all programming is done by Kenyans, possibly by local consultants; (4) good microcomputer operators are retained by the civil services; and (5) maintenance and supplies are provided for in the budget (Pinckney, 1984).

Agricultural Research

Another microcomputer application occurs in agricultural research. The work of the agricultural researcher, like that of the agricultural institution or project manager, is information-intensive. Success is often dependent on the existence of accurate databases and powerful data processing tools. Before the advent of microcomputers, researchers in developing countries faced the choice of either processing and computing their data manually or using a centrally located mainframe computer. The introduction of the microcomputer into developing countries has simplified the work of their agricultural researchers. Microcomputers are now being used to perform functions ranging from the collection and processing of agricultural survey data to the statistical analysis of experimental field data and the computation of least-cost feed-grain mix.

One of the primary benefits of the microcomputer in developing countries has been in easing the process of conducting surveys and analyzing survey data. This process is vital for accurate agricultural research databases and for the best design of agricultural projects. Among the many advantages of using the microcomputers for agricultural survey processing are:

- Microcomputers are portable and durable, when compared with mainframe or minicomputers. They are capable of storing large amounts of data; therefore collection activities and data processing activities can be in closer proximity.
- Software for microcomputers is considered to be more "user-friendly," requiring less training than the software on mainframe computers.
- Data processed on microcomputers can easily be aggregated for further processing on a larger computer.

These advantages have convinced agricultural researchers in a number of countries that microcomputers are appropriate tools for these tasks. Table 3-1 lists example of microcomputer use in the processing of statistical survey analysis and database composition.

Statistical Analysis

One important component of any program to improve agricultural research systems in the developing world is the design and analysis of research trials and farmer characterization surveys. Data collection, management, and analysis need to be organized to avoid typical data handling problems and to provide step-by-step procedures to solve agricultural problems. Many national research programs and agricultural universities already have microcomputers because of their relatively low cost. However, there are few, if any, software packages available that can satisfy all the specific requirements of agricultural research.

Olvind Nissen, Agricultural University of Norway, in cooperation with scientists at Michigan State University, has developed a microcomputer software package, MSTAT, to facilitate agricultural research. Already in use in several countries, it helps researchers design and manage complex experiments, process survey data, and analyze with greater ease and speed. It provides timely

Table 3-1   Microcomputers   for   Census   and   Agricultural
            Research in Developing Countries

Country	Application	Microcomputer
Bolivia	Agricultural Statistical Analysis	ONTEL
Cape Verde	National Census	Billings/Onyx
Cook Islands	Population Statistics	ALTOS
Ecuador	National Census	ALPHA
Egypt	National Census	
Francophone Africa	Collection of Agriculture Field Data	TRS-80
India	National Sample Survey	HCL-8C
Indonesia	Health and Agricultural Survey	Obsborne
Jamaica	Crop Production Survey Pig Census	North-Star
Mali	Infant Mortality	IBM/PC
Mauritania	Fertility Survey	ALTOS
Nigeria	Agriculture Survey	Apple
Puerto Rico	Economic Census	Apple
Sierra Leone	Agriculture Remote Sensing	North-Star
Singapore	Statistical Analysis	Apple
Tunisia	Rural Development Information System	Apple

Source: Diskin, et al. "Considerations for Use of Microcomputers
in Developing Country Statistical Offices," International Statistical Programs Center, U.S. Bureau of the Census, Department of
Commerce, Washington, D.C. 1983.

research results and thereby facilitates the generation of new and appropriate technologies.

The main program of MSTAT and its 50 subprograms occupy approximately 220K of disk space, and they come in two separate program disks. Program use of resident microcomputer memory can vary from 40K to 60K depending on the particular subprogram being executed. The main program and the subprograms are written in MBASIC which means that a user can alter them for special applications. Table 3-2 shows several different aspects of agricultural research where MSTAT programs can be used.

The program is written to be easy to use; it is flexible and operates in a dynamic and interactive mode. Some statistical knowledge and programming capabilities are required to use the program properly. Although originally designed for plant breeding and agronomic uses, it is directly usable for laboratory experiments as well as socioeconomic survey data analysis. Some present MSTAT capabilities are:

- Experimental design generation
- Field map printing for laying out field experiments
- Label printing for above designs giving planting, field stake, sample, and harvest labels
- Field notebook printing for the above designs to facilitate data collection
- Data manipulation capabilities
- Data analysis using basis statistical techniques.

Other Uses of the Microcomputer and Microprocessor in Agriculture

The use of microcomputers for agricultural research, institution management, and project management by no means exhausts the potential of the microcomputer for the agricultural sector in developing countries. It is helpful to investigate the use of microcomputers in the industrialized (and newly industrialized) nations to understand their full potential for agriculture in developing countries. In addition to the uses described in the next chapters, microcomputers and microprocessors are being used as the key link in automated processes, such as irrigation control and food processing control. Microcomputers are also widely used in agricultural research for the construction of agricultural models and simulations. Finally, microcomputers are being used directly on the farm. Such "on-farm" computer systems can perform a variety of functions ranging from intelligent terminal communication with national agricultural databases to running farmer-written software for specialized applications. A brief overview of these applications follows.

Table 3-2   Areas Where MSTAT Subprograms Can Be Used in Agricultural Research

Area	Steps Involved	MSTAT Subprogram
Characterization farmer and farming identification, and recommendation	Survey data, data analysis, frequency tables, mean values correlation, regressions, multiple regressions, economic analysis	FORMREAD REGR, COR, MUITIREG, FREQ, ECON
Testing and verifying technologies and hypotheses	Design experiments	EXPPLAN, VARPLAN
Conducting and managing experiments and data	Print field maps, fieldbooks, and labels	EXPBOOK, EXPMAP
Analyzing experiments	Analyze experimental data, analysis of variance routines, regression correlations, multiple regression, economic analysis	ANOVA-1, ANOVA-2, ANOVALAT COR, FACTOR, NONORTHO, REGR, CALC, HIERARCH, MULTIREG, GROUPIT, TABTRANS, STAT, ECON
Report writing and recommendations	(In combination with a word processor) Print tables, means, histograms	FREQ, MEAN TOTEMP, FROMTEMP,

Irrigation Control

Successful development of agricultural systems often requires a precise control over inputs, including water. A microprocessor can be used as the central element of an automated irrigation control system. The microprocessor is attached to electronic probes that measure environmental parameters such as soil type, humidity, and weather conditions. Using this information, the microprocessor can regulate the operation of an irrigation system. Thailand is currently considering the commercial introduction of advanced modular technology for a low-cost field-level irrigation system in its northeast region (Bhalla et al., 1984).

Computer Control of Food Processing

Food storage and processing is another agricultural function appropriate for automation . Microprocessors and numerically controlled machines (including robots) are used to control and optimize food production processes in sugar factories (Cuba), rice-patty makers (Japan), canning factories (Hungary), and stockyards (USA). Microprocessors are also used to monitor food storage facilities and to control the environment in greenhouses and dairy cattle barns. In India, a microprocessor has been developed for measuring and recording the fat content of milk. This device was produced as a result of research and development in a major cooperative society in western India. This decentralized cooperative collects milk from members twice daily. Records are kept of the fat content of each sample, since prices for the milk vary with the fat content. This replaces the cumbersome previous method of recording this data by hand. In Thailand, a rice miller has bought a new rice-sorting machine with eleven microprocessors. The machine sorts out the grains that are diseased or stained, and thus improves the quality of the rice (Bhalla et al., 1984).

Modelling and Simulation

Computer models and simulations are sets of interrelated mathematical equations designed to recreate a real-world process. These models and simulations are powerful analytical tools for agricultural researchers because they can test alternative strategies rapidly. Although complex models and simulations have traditionally been built and used with mainframe computers, they are beginning to be developed and applied on microcomputers as well. Simulations have been built to model the epidemiology of certain animal diseases, the production performance of dairy herds, the relationship between agricultural prices and income distribution, the management of sugarcane enterprises, and demographic models of livestock.

The "On-Farm" Computer System

In the United States, which has a history of capital-intensive farming, the microcomputer has become a valuable device on farms. Microcomputers are being used for generic functions, like calculating budgets, and a host of user-specific functions like calculating optimal land use for a specific farm. Many farmers write their own microcomputer applications that they then sell or trade with other farmers. The Tandy-Radio Shack company maintains a current listing of more than 300 farmer-written software packages that are available for sale and exchange.

Another function for the on-farm computer is to provide communication capabilities with national agricultural databases and to establish electronic mail systems among farmer interest groups. Using a microcomputer and a modem (a device that modulates and demodulates electronic signals to send information over the phone lines), farmers can link up with a wide range of computer networks. These networks provide information on weather conditions and forecasts, commodity prices, and current market prices for livestock.

SELECTED EXAMPLES

Agricultural Management, Kenya

In 1981, a study was carried out at Stanford University's Food Research Institute to investigate the application of microcomputers to improve information management in the Kenyan Ministries of Agriculture and Livestock Development (Pinckney et al., 1982, 1984). The study, jointly sponsored by the USAID mission and the Harvard Institute of International Development's team in Kenya, explored the use of microcomputers for:

- Improving the ability to evaluate the consistency and implications of crop forecasts
- Increasing the capacity to assess the status of food security within the country
- Providing more systematic and timely data on budget expenditure, personnel, vehicle control, and payroll administration

- Developing increased capacity to provide information on the status of project implementation
- Developing a word processing capability that would reduce the time required to prepare project feasibility studies.

The most immediate need identified for microcomputer use was in the area of budget and finance for the ministries. It was decided to use an Apple II microcomputer with 64K of RAM and a CP/M operating system card. The advantages of using the Apple in Kenya were: (1) the large amount of software available, particularly when the CP/M card is installed; (2) the Apple's high level of resistance to fluctuations in power supply; and (3) the presence of other similar systems in Kenya.

The following applications to improve the management performance of the Ministries of Agriculture and Livestock Development were recommended:

- Improving the Expenditure Reporting System. Before the implementation of the microcomputer, expenditure reporting suffered from a number of major problems: it was unclear, untimely, inaccurate, and required a duplication of effort for different users. A Visicalc template was constructed on the microcomputer to standardize the expenditure reporting system and to make it more timely, accurate, and easier to compile. Budgeted expenditures could be compared with actual expenditures.
- Budget Overexpenditure Decision Making System. A very creative use of the microcomputer was in the construction of monthly budget statements for the ministries. The microcomputer became a part of the budget process when it was used on a number of occasions as a tool to aid in budget planning at committee meetings. By using a model and testing alternative strategies on the models, planners could instantly see the result of budget decisions. The discussion proceeded on an item by item, section head by section head basis, with further expenditures being frozen in some places and allocations being reduced in others.
- Budget Estimates System. A similar application was implemented for the estimation of future budgets. In the past there had been no useful summary of the estimate process for the ministries' forward budgets. The typical draft covered more than 100 pages of stenciled copy, making it an unwieldy management tool. Using the microcomputer, managers were able to devise a five-page summary format for future budgets. The format presented basic financial data for each subsection and

included budget authorization, expenditure, and forward budget figures. The simplicity and brevity of the summarized budgets facilitated the managers' efforts at controlling spending.

● External Aid Claims System A central operating weakness for the Ministries of Agriculture and Livestock Development had been the inability to track outstanding reimbursements owed by foreign donor institutions. (For example, in 1981 the Ministry of Agriculture had 51 donor-assisted projects in which 64 percent of the costs were to be paid by the donors; less than half of those billable expenditures had been claimed).

Agricultural Production, Portugal

The introduction of microcomputer technology for use in the Portuguese Agricultural Development Project (referred to as PROCALFER) came as part of a USAID program begun in 1979 and carried out with the help of the United States Department of Agriculture (Ingle and Connerly, 1984). The initial goals of the agriculture project included:

● A national limestone application program to address the problem of high soil acidity
● An accelerated forage and pasture expansion program to improve soil quality and provide feed for livestock
● Reorientation and improvement of the national research and extension institutions to better meet farmers' needs.

Apple microcomputers were introduced in the PROCALFER project in 1983 to enhance managers' capability for project design and implementation. As in the Kenya Ministry of Agriculture, the microcomputer has had most value in the area of management improvement, especially in relating to the performance of cost estimation and budgeting tasks.

Other proposed applications include:

● scheduling time for implementation planning and depicting dependency relationships among key activities
● monitoring national-level program activities and accomplishments (for example, maintaining records of limestone production and distribution, credit applications and loans, and seed distribution)

- monitoring regional-level program activities and accomplishments (for example, planned versus actual project outputs on a periodic schedule)
- maintaining program budgets, financial accounts, and records for funding agencies
- scheduling and monitoring USDA inputs and activities (for example, scopes of work for technical assistance teams, training requests and placements, and procurements)
- coordinating group records and documentation and reporting (such as monthly reports on PROCALFER status to officials).

Some technical applications of microcomputers in this project include:

- Policy Analysis Studies. The Policy Study Team indicated that access to a small computer would be valuable in their statistical and modelling work over the next few years.
- Limestone Application Research. Members of the Coordinating Group indicated that a microcomputer would facilitate their ongoing research, since all calculations were currently being done by hand.
- Farming Systems. This planning team indicated that a microcomputer would be invaluable for data collection and analysis if their plans were accepted by the ministry.

Microcomputers have helped to redefine relationships between regional and national authorities concerned with the project in several ways:

- Because they are a "seductive" technology, microcomputers have encouraged the regions to participate in the management efforts of the PROCALFER project.
- Microcomputers decrease the drudgery and increase the accuracy of operating information systems. The ability to construct a timely and accurate database transforms the rivalry between the two management factions from one based on territoriality and personal antagonism to one more constructively based on a debate over budget figures and fiscal planning.
- Microcomputers established information system boundaries. While microcomputers encourage regional offices to participate in the project, the central coordinating group has control over the timing and implementation of microcomputer use in PROCALFER. This control allows for the establishment of an integrated information system, shaped by national as well as regional priorities.

Agricultural and Rural Development, Nepal

Microcomputers have been introduced in over 19 development projects in Nepal, serving a variety of management functions for agricultural and rural development projects. The establishment of a national users' group is seen as a crucial part of the long-term acceptance and profitable use of microcomputers (Sharma, 1984). Table 3-3 summarizes some typical applications of the microcomputers in Nepal.

Table 3-3   Applications of Microcomputers in Nepal

Program	Application
Family Planning and Maternal Child Health Project	Monthly reporting system of 10-20 services provided by 1,500 family planning clinics nationwide.
	Monitoring of target versus actual reports.
Health Planning Unit	Sales monitoring.
	Financial planning.
Rapid Development Project	Monthly financial analysis of 21 line items in 54 offices.
Community Forest Development Project	Socioeconomic survey data analysis.
	Projected versus actual figures on: finances, nurseries constructed, forests planted and transferred to village control.
Resource Conservation and Utilization Project	Procurement tracking system.
	Payroll and personnel records.

Farm Management Surveys, Nigeria

The Agricultural Projects Monitoring, Evaluation, and Planning Unit (APMEPU) is a part of the Nigerian government created in 1975 to monitor and evaluate agricultural development projects throughout Nigeria, (Bennett et al., 1982). The standard evaluation is based on a set of statistical surveys carried out on a sample of smallholder farmers for each of the projects. In the 1982-1983 cropping season, surveys were carried out in 15 agricultural zones.

Processing of this survey data has historically been constrained by the limitations of the mainframe computer assigned to the task. The CDC Cyber 72 mainframe, although adequate in terms of processing power, suffered from power shortages, voltage spikes, and poor maintenance. As a result, there was no substantive analysis of APMEPU data collected from the field for the first three agricultural projects.

The introduction of Apple II microcomputers remedied this situation in APMEPU field offices. Data from the survey questionnaires are entered and stored on diskettes. The data diskettes are then forwarded to the APMEPU headquarters, where a Data General minicomputer is used for statistical analysis.

The following four surveys have been or are in the process of being carried out on the Apple:

- Agronomic Analysis Survey. An average survey consists of 1,200 field and plot records. The Field and Plot Report and Household Report are the most important.
- Clearline Survey. This is the latest of a series of farm management surveys, which collects weekly data on items such as crops, fertilizers, and labor utilization over a one-year period. Nearly 6,000 survey forms are involved.
- Price Survey. This survey takes a periodic look at current market-basket price. It is the simplest of the surveys undertaken.
- Baseline Survey. This survey attempts to detail farm size, ownership, household composition, off-farm activities, cropping patterns, farm input, livestock holdings, and asset ownership for a large number of households.

Food Crop Research in West Java, Indonesia

The Agency of Agricultural Research and Development (AARD) in Indonesia is committed to provide a modern computing facility in Jakarta. Various research institutes under its organization are widely spread across the country, however, posing limitations to the development of an effective computing network. Communication has been a problem, despite the fact that Indonesia is one of the few countries that has operated its own domestic satellite for some time (Tohar, 1984).

In 1980, a TRS-80 microcomputer was introduced at the Sukamandi Food Crops Institute, and research professionals and technicians have been gaining valuable microcomputer experience. The microcomputer has proved to be an indispensable tool for the Institute. Researchers have succeeded in modifying and developing certain programs to meet their own needs. Specialized programs now include the following modules:

- A data management system that creates data files to be used as input to the analysis of variance. Data can be modified or printed before the analysis.
- Completely randomized design techiques for selecting experimental plots.

Future goals include the establishment of a microcomputer network and microcomputer links with the AARD centrally located mainframe computer.

Least-Cost Feed Formulation, Taiwan

Using a microcomputer for calculating feed formulation is similar to traditional hand calculating methods; the data necessary for the calculation and the results are usually the same. When the best feed formulation is arrived at, it is also necessary to account for other factors associated with the feed's effects, including cost, palatability, toxicity, and effect on development. In research carried out at the National PingTung Institute of Agriculture in Taiwan, the microcomputer is being used to make these computations (Hsieh, 1984).

In this context, the microcomputer has distinct advantages over the traditional methods of calculation:

- Clearly, a greater volume of data can be handled when using the microcomputer than is possible in hand calculating.
- Microcomputers can improve the accuracy of the result. The most important aspect of this operation is the maintenance of an accurate data file for feed composition. Using a microcomputer allows one to alter the configuration of the data file with tremendous ease. This

flexibility greatly facilitates the calculation of the formula.

● Microcomputers make the task of computing multivariable regressions much easier than with traditional hand methods. The calculation of a least-cost feed formulation includes nutrient requirements, palatability, taste, an unknown growth-promotion factor, pigment, and feeds resources availability.

The objective of using the microcomputer is to arrive at the least cost combination of ingredients that also satisfies all the necessary requirements. A standard linear programming methodology is used to calculate the least-cost feed ration. The programming algorithm is illustrated in Figure 3-1.

As shown in the flowchart, the first step in the process is the construction of a data file. The next step includes choosing and putting in feed items, feed prices, ingredient-tolerance levels, and prices. Analysis of preoptimization involves excluding unusual or unfeasible limitations. The program devised uses the general simplex method for linear programming of the objective function. Output includes a rations ratio, a nutrient ratio, and sensitivity analysis. Sensitivity analysis is an important feature because it provides the shadow prices and opportunity costs of inputs. A shadow price allows one to account for nonprice limitations on an input. For instance, soya may be a desirable feed input and available at a subsidized price but in limited quantities. The shadow price reflects the true price of the input, which takes its scarcity into account.

Conclusion

These case studies point to a variety of applications that are currently under way. They represent only the tip of the iceberg, however. Since this volume was written, many new applications in agriculture have emerged in the areas of expert systems, farm equipment and irrigation control and remote sensing applications for crop monitoring, in particular.

These topics as well as advanced technologies emerging in other sectors, are the subject of the third symposium in this series.

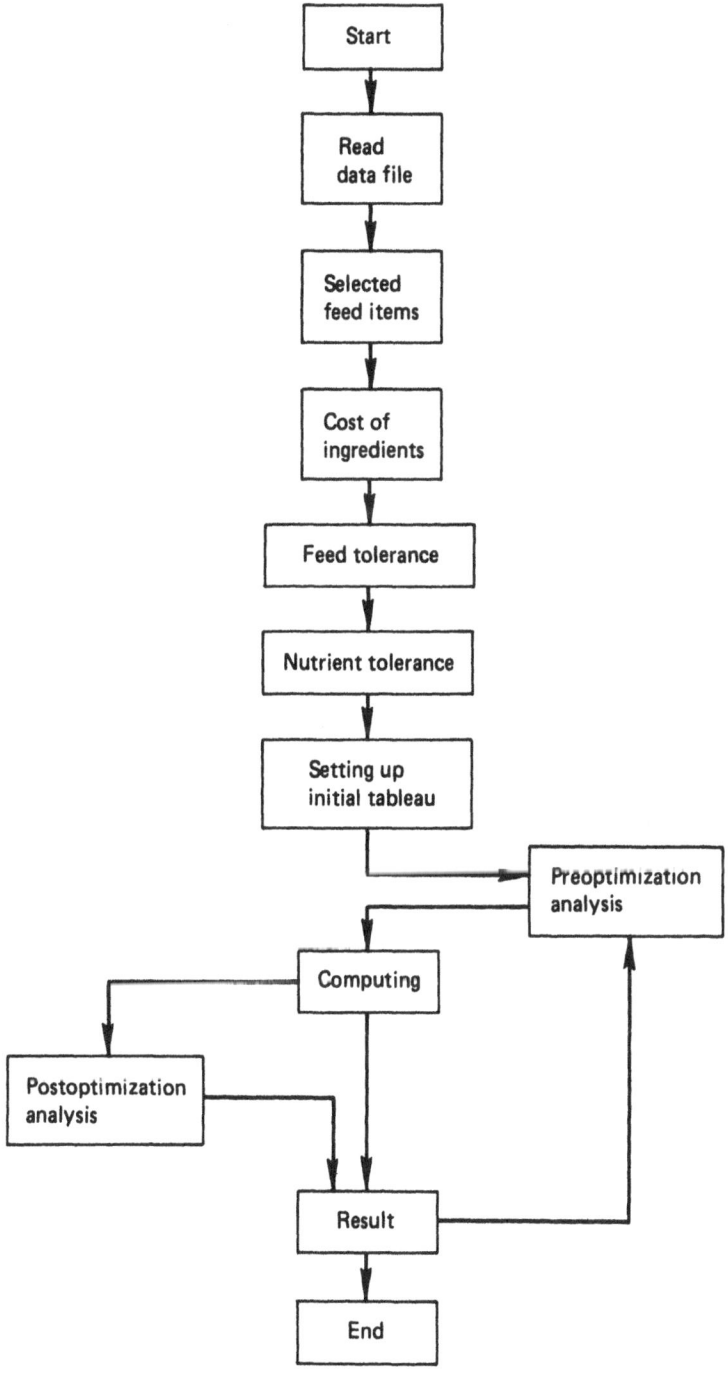

Figure 3-1   Least-cost feed programming algorithm

4

Applications in Health

INTRODUCTION

The purpose of this chapter is to describe the diverse uses for microcomputers in the promotion of improved health care in developing countries. It does not cover all possible uses of the microcomputer in the health sector, nor does it evaluate the success of the reported projects. Rather, it attempts to give the reader an overview of the many ways these systems can be used at varying levels of sophistication, from simple storage and retrieval of information to complex modelling. Examples of projects using microcomputers in the field at intermediate levels (such as regional or district offices) and in central ministries are reported as well.

Health includes not only the many facets of medical care and treatment, but also nutrition, health education, epidemiology, demography and population, and environmental science and engineering. Rather than classify the uses described by discipline, the approach adopted here is to categorize applications according to technical levels of microcomputer use, or what is often referred to as the degree of sophistication or "smartness." Almost every health discipline involves use of computers at each of the levels described.

The more sophisticated levels depend on preliminary simpler analysis. First the data are collected and stored, then subjected to statistical analysis. Finally, based on this information, a model can be created. The model, once constructed, can be subjected to changes in individual variables to study their impact. In this way, a large number of alternatives can be tested in a short time.

The first category to be discussed is Database Management, in which the computer is used for storing and retrieving information. The next level of sophistication goes beyond mere retrieval of data to include Statistical Analysis. The third category involves use of the microcomputer for <u>Modelling and Simulation</u>, which play an important role in health planning and decision making. The microcomputer also has a role to play in <u>Ancillary Functions</u> in connection with other equipment such as the complex diagnostic tools used in hospitals that require the processing and display of information.

Case studies will illustrate applications in each of these categories.

Databases for Health-Related Information

References on the use of microcomputers in the health field can be found in three computerized databases: Medline 1979, Compendex (an engineering index), and Exerpta Medica. In addition, several conferences covering health and microcomputers in developing countries have also been held, notably the National Council for International Health (NCIH) conference on Computer Technology and International Health, held in Washington, D.C. in January 1984. These conferences are being held often and they are an important source of information. Some recent examples include:

- World Congress on Medical Information Systems and Developing Countries, Mexico City, February 7-12, 1982, sponsored by WHO, PAHO, and the International Medical Informatics Association
- Fourth World Conference on Medical Informatics, Amsterdam, August 22-27, 1983, sponsored by the International Medical Informatics Association
- Eighth Annual Symposium on Computer Applications in Medical Care, Washington, D.C., November 4-7, 1984
- Fourth Brazilian Workshop on Microelectronics, Campinas, Sao Paulo, February 21-March 4, 1983, with a portion devoted to "Biomedical Applications of Integrated Circuits."

An example of information shared across national borders is POPLINE, the world's largest bibliographic population database; it has over 2,000 users in the United States as well as 14 centers abroad with direct access. Johns Hopkins University's Population Information Program does searches for other users overseas, and it plans to expand the database to microcomputers to make the information more readily available around the world. Consideration is also being given to producing a POPLINE tape of French-language material on the Sahel nations for the Reseau Sahelien d'Information et de Documentation Scientifiques et Techniques

(Sahel Network of Scientific and Technical Information and Documentation).

AID, which contracted with Johns Hopkins to create POPLINE, has also contracted with the International Science and Technology Institute (ISTI) to develop a health projects database for use on microcomputers. AID had earlier commissioned a database for water supply and sanitation projects, under the Water and Sanitation for Health Project (WASH).

Volunteers in Technical Assistance (VITA), a nonprofit organization, also has an interest in computerization. Many of their projects are currently abstracted for the DEVELOP database operated by Control Data Corporation. These involve appropriate technologies of all kinds, including water and sanitation, food processing, prosthetic devices, and other areas of health. VITA has been involved for several years with satellite conferencing and plans to tie that technology into microcomputers in the next few years. Its purpose is to transfer written information from one microcomputer to another. VITA uses microcomputers in its home office for administrative records, bibliographic references, and to store the resumes of 4,000 volunteers, who can be readily identified by skill area, language proficiency, and experience.

APPLICATIONS REVIEW

Database Management

Field Information Management

Storage and retrieval of information is probably the most basic use that can be made of the microcomputer and possibly the one applicable to the largest number of users. In the health sector, computers perform three main data processing functions. First, they allow fast and accurate storage and retrieval of data by category for example, the number of patients from a certain geographical area or ethnic group, the average age of patients with a certain disease, a listing of all homes that use well water). Second, data retrieved can be presented in various formats to highlight different relationships. Finally, computerized information offers a powerful decision making tool by allowing the processing of large amounts of data; the information can then be extracted from stored data and projected into the future for planning.

The recent development of relatively inexpensive and extremely effective interventions, such as vaccines against diphtheria and poliomyelitis and oral rehydration therapy for diarrheal disease, have emphasized the need for good health information systems. To implement a health information system, data must be collected, stored, processed, analyzed, and interpreted; the resulting information must be disseminated in a timely manner, and feedback is necessary.

Data are usually collected at the local level, usually by health care staff. As the data from many local health centers are aggregated, the need for analysis, for dissemination, and for timely response creates a demand for computer processing. In many countries, a central computer is used, but delays arise in communicating the data to the central collector and in disseminating results. When the central computer is overloaded, as is often the case, the delays defeat the purpose of the information system. Several projects have used microcomputers to enter and tabulate field data before sending them to a mainframe computer for detailed analysis.

Many experts have therefore concluded that data processing and dissemination should be done from an intermediate point between the national or central level and the local health care units. At this intermediate level, the microcomputer provides a useful tool, and, being flexible, it can be adapted to suit the needs of the user.

At the local level, a modest effort is being carried out by A.P.R. Aluwihare of the Department of Surgery, University of Peradeniya, Sri Lanka, who uses a low-cost personal microcomputer (in the $100 range) to organize patient data. He monitors the long waiting list of patients with different disorders and stores information on predicted versus confirmed diagnoses; he also compares predicted and confirmed lengths of hospital stays. He follows patients to ensure that they do not get "lost" if they fail to arrange for the surgery that has been recommended. He also uses the microcomputer to prepare examinations for medical students by storing previous questions according to their characteristics. He is able to use the computer to retrieve the questions by category to satisfy different objectives for the examinations. He is also using the computer as a teaching tool to assist Sri Lankan medical students to improve their spelling of technical and medical terms in English. His eclectic approach clearly demonstrates how a medical professional can use the microcomputer to help minimize the time given over to clerical tasks, (Aluwihare, 1984)

In Thailand, (Boonthai, 1984) a similar need exists to collect, analyze, and disseminate information about health problems and policies. Health care providers from the village "tambol" doctor to the Ministry of Public Health (MOPH) need to be linked so that information can flow smoothly up and down the chain. Figure 4-1 shows the number of groups that have been included in the Thai system.

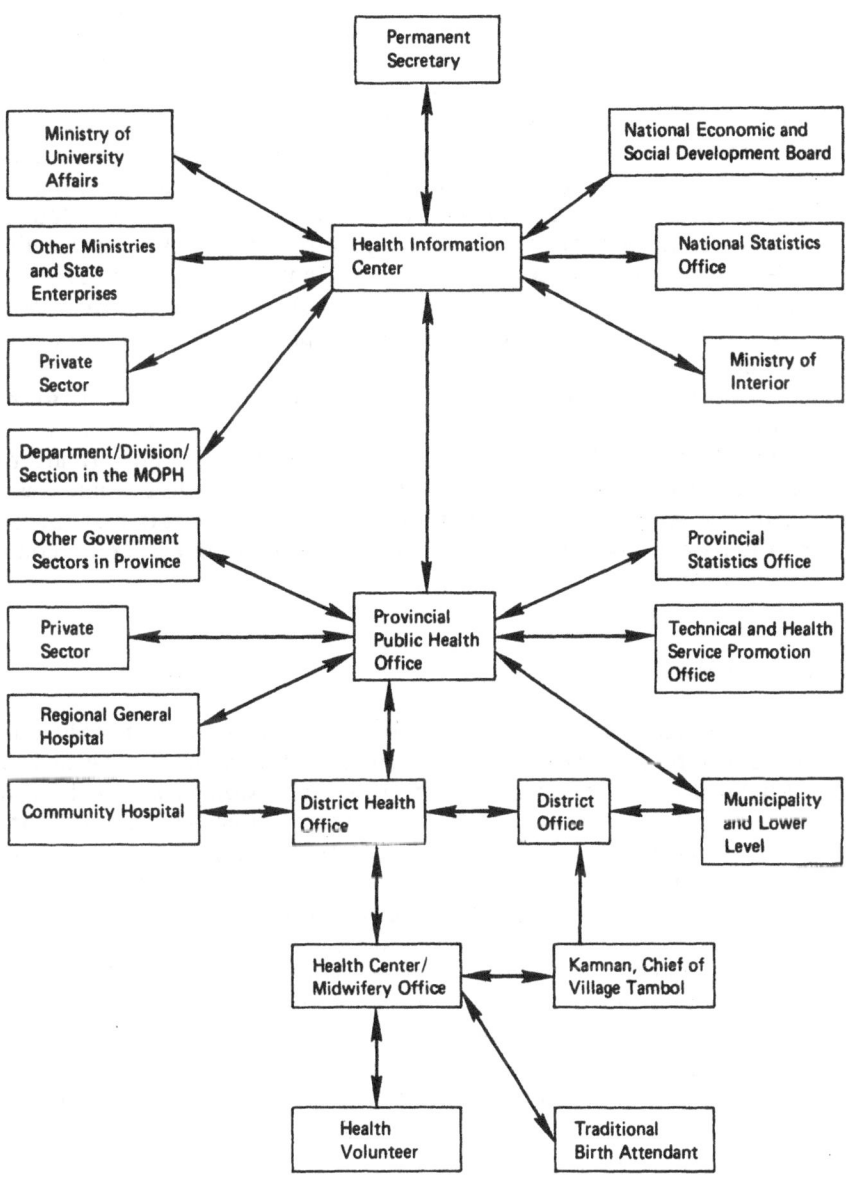

Figure 4-1 Organization of the Thai public health system

Health posts at the village level, namely the Health Center and Midwifery Center, collect data and fill in the record form designed by the MOPH. The basic data are collected from health volunteers, traditional birth attendants, the chief village doctor, village headmen, and from the activities of the health posts. The data are then sent to the district health officers, after which they are compiled and forwarded each month to the province in report form. The provincial chief medical officer general uses the available data for monitoring health activities. He may study some specific data for solving technical and managerial problems. Finally, aggregate data are filed in another report form and forwarded to the central level every three months. In poverty-striken areas, the health information system is monitored more closely than average because the problems in these areas are urgent and require immediate attention. The progress of every project must be reported monthly.

Because compiling these reports by hand is slow, and information relevant to smaller aggregations is lost, the government is studying the use of microcomputers to collect and transmit information. Efforts are underway to devise a system that would take account of existing microcomputers in use in hospitals and larger health offices and, as far as possible, make a new system compatible with them.

Hospital Management

Computerization is already quite common in hospitals and clinics where personnel profiles, equipment and drug inventories, and payroll data are stored in the computer. Computer use is also appropriate for government health offices, collecting and presenting nationwide data. For example, Olson (1984b,c) reports on the use of two microcomputers in Lesotho's Ministry of Health Planning Unit. These machines (with an uninterruptible battery power supply) have been in use since August 1982, giving support to finance, personnel, statistics, and nutritional services. The computers were originally meant to handle the national health statistics that were swamping the ministry. Data tabulations (of services provided, disease frequencies, length of hospital stay) were previously done by hand, and reports could take years to complete. Now, however, national, local, and district data are available the following quarter.

In addition, the computers are used to keep equipment inventory; updating the records takes only a few minutes. Botswana and Swaziland have become interested in Lesotho's data processing system, and Lesotho has begun to provide them with technical assistance.

CBR Associates of North Carolina is one of many organizations developing software packages for use with microcomputers in United States hospitals. CBR visited several small hospitals in the U.S. to determine the areas that could best be addressed with microcomputers. Out of that project, CBR has developed several programs for the health care field, incorporating commercially available software packages that are relatively "user-friendly." With modifications to suit local conditions, the four management systems described below would be applicable to developing countries as well:

Facility Utilization and Scheduling Systems. Because of high hospital costs, cost control is essential. For the benefit of health organization and health insurance companies, cost reviews can be established to help estimate reimbursable medical care or service costs.

Monitoring systems have been developed for scheduling medical personnel, for monitoring the inpatient occupancy rate, and for controlling emergency room and operating room services.

In hospitals with high rates of elective cases, the waiting list for medical procedures can be scheduled by microcomputer. Length of stay for individual nonemergency cases can be computed as base line data from diagnosis of diseases, treatment schedule, age, and sex. The number of patients discharged in a day can then be determined, together with the number of vacant beds.

Emergency room outpatient registration software for entry of information on patients admitted to the emergency room or to outpatient clinics has been developed. Registration of patients can be simplified on subsequent visits, and reports made on the types of patients being treated and the number seen each day or by each physician.

Financial Information Systems. The objective of a financial information system is to provide data to aid in decision making by assessing hospital services and monitoring budget and expense. Microcomputers can be applied to various hospital accounting subsystems, including payroll, cost and schedule of supplies, accounts payable, and financial status.

Materials and Facilities Management Systems. Microcomputers can be useful in material and facilities management in such areas as procurement and effective utilization of facilities and equipment. Some areas in which they are already proving effective are in drug inventory control, menus for meals and nutrition, monitoring and control of repairs and maintenance, and inventory procedures that track equipment and other items by location and report minimum and maximum levels. Energy use in the hospital can also be monitored by packaged programs, and can result in improved control and economies in energy utilization.

Finally, the construction of new hospital sections can be monitored by computer through project management. Package programs such as Program Evaluation and Review Technique (PERT) can be applied.

Personnel Data Systems. Personnel data systems store general data on personnel, including employment history, evaluation summary, insurance information, health information, and educational history. Other packages are available that include medical staff profiles (such as information on certification, licensing, and membership in professional societies); data on liability insurance and any past claims brought against each physician; data on committee membership, and a record of each physician's continuing education credits. Nurse scheduling programs are available that keep a master schedule of nurses by skill level, unit, and shift over a period of one month. Schedule changes can be made quickly, holiday and vacation time can be tracked, and up to 350 individuals can be scheduled at once. The scheduling (a modelling use of the computer, rather than just a database) can be done with commercial software programs such as Lotus 1-2-3.

Demographics

Demography in developing countries has only recently passed the point where the foreign expert devises the survey, conceives the coding scheme, supervises data entry, and analyzes the results. The computer analysis was often done in his home country, and results seldom filtered back to the developing country for action. Studies by the Population Studies Center at the University of Pennsylvania indicate Third World demographers returning to their countries have often been thwarted by a lack of computer resources (Strong, 1983). With the availability of small, inexpensive, and reliable microcomputers, this situation has now changed. The microcomputer allows researchers to bypass the inefficient national mainframe systems (which may be bogged down in government payrolls and international trade statistics) and to allow them to engage in autonomous research. For the present, at least, decentralization is the most promising model for that research.

With the present cost of microcomputers, the solution of one computer per survey, which was applied in Bamako, Mali, is viable. The young professional in an isolated location can now rely on a tool that will greatly reinforce his potential usefulness. The major bottleneck, slowing the diffusion of this new technology among demographers, is the lack of software (the programs implementing these quantitative techniques), written for microcomputers.

An ideal way to teach demographers how to use the software for a particular program is to let them learn it during their graduate education. The Population Studies Center is planning to establish a microcomputer laboratory for just this purpose. A second-best alternative would be to have demographers receiving the software attend training workshops for perhaps two or three weeks, during which they could become familiar with both the microcomputer and the programs. The workshop should of course be located in the user's region (for example, for Africa at the Sahel Institute). A less-desirable alternative would be to send consultants to teach in each country. Unless a number of people can be taught at once and unless they can devote full time to the course, this method can be more expensive and less effective. The least-attractive alternative is to simply ship the hardware and software to the researcher without any kind of training or consulting program.

Statistical Analysis

Analysis of Survey Data

In addition to storing and retrieving information, computers can be used to perform statistical analyses of data (that is, calculating means, standard deviations, and correlations). In the past, after data were collected in the field (on nutrition, latrine use, or whatever), they had to be laboriously transcribed onto computer coding cards and then sent to the mainframe, which was often overseas. With microcomputers, the data can be entered directly in the field, and inconsistencies can be identified and corrected. The results are available immediately and can be used to revise and improve the remaining data collection efforts.

Health and Socioeconomic Indicators

There are many reported uses of microcomputers in the analysis of survey data. Under an AID grant, for example, Fernando and Sandra C. Bertolli (1981) set up a statistical analysis program in Morocco in cooperation with the Ministry of Health. They used the Interactive Statistical Inquiry System (ISIS) on an Apple II microcomputer to create and maintain on-line data files. The core of the database is the epidemiological data that have been collected by province since 1969. The primary advantage of the computerized system is the speed with which it manipulated, displayed, and printed information, leaving highly trained personnel more time for analysis. The system is expected to serve two main functions: monitoring and evaluation. The monitoring function will be served by analyzing incoming data in a timely manner, so that trends of incidence rates of the monitored diseases can be discerned. Given dependable data, the system can be used to provide warning of a potential epidemic outbreak.

Maternity Care Information System

Research Triangle Institute in North Carolina is one of many organizations that has been involved in promoting the use of microcomputers for statistical analysis. They have provided hardware, software, and training to the Department of Statistics in Rwanda, the Institute of Statistics and Demography in Burkina Faso, and the Ministry of Finance in Tunisia, as well as to other government ministries outside the health sector.

Family Health International (FHI) developed a maternity care monitoring system at the request of physicians who wanted a "maternity survey" as a tool for improvement in maternal and perinatal health (Whitehorne and Trottier, 1984). The system was designed to identify and monitor areas where specific interventions might be needed.

The maternity care monitoring system is based on a single-sheet, precoded questionnaire, known as the Maternity Record, used to record data on deliveries in hospitals, maternity centers, and other health service facilities. Information recorded at the time of delivery includes sociodemographic characteristics, obstetric history, contraceptive practices, antenatal care, management of labor and delivery, and maternal and perinatal outcomes. A "Maternity Record Summary," a shorter version of the original questionnaire, was also developed, leaving out those questions less relevant for smaller, less well-equipped and well-staffed centers.

Initial pretesting of the system was a cooperative effort between FHI and the International Federation of Gynecology and Obstetrics (FIGO), involving more than 20 hospitals in almost as many countries. After pretesting, three full-scale test sites were chosen in Indonesia, Thailand, and Tunisia. Initially, this system was used in major referral centers and teaching hospitals, where it continues to serve as an important source of information to meet a variety of needs.

Recognizing that in many countries the vast majority of women deliver their babies at home assisted by a traditional birth attendant, FHI has also supported projects aimed at obtaining information on noninstitutional deliveries.

The Maternity Record collects data on obstetric care that can be analyzed and reported quickly to participating institutions so that health professionals will be better able to manage high-risk mothers and infants. The resulting data are useful to clinicians, program administrators, health planners, and researchers as a comprehensive summary of events in the wards and at health posts and even among children delivered at home. Analyses based on the data have contributed significantly to the literature on pregnancy-related care in the developing world. Among the topics that have been addressed are high-risk pregnancies, cesarean deliveries, determinants and consequences of birth spacing, and evaluations of family planning services offered immediately after childbirth.

Maternity data collected through this program have been used in several countries to generate country-specific pregnancy risk indexes that health care providers can use to identify women needing special attention and services to ensure a favorable pregnancy outcome.

Maternity data from several countries in sub-Saharan Africa have been used to demonstrate the effect of prolonged breast feeding on birth intervals; the data have shown that, in some countries, the trend toward shortening breast-feeding duration, without replacing the lost birth-spacing effect with modern contraception, results in shorter birth intervals, with higher risks for mothers and infants, and an anticipated higher total fertility rate.

The maternity care monitoring system, combined with micro-computer technology, provides an invaluable tool for health-oriented research. Access to information for improving health care is now available almost immediately. In keeping with the total systems approach aimed at institution building, software packages have been designed and implemented to complement the original programs. Complete systems for studies of different contraceptives were designed following the maternity care mode. These systems include capture, entry, and verification of data; internal range and consistency checks; error reporting; and a number of standard administrative and clinical routines to be used for statistical analysis.

Population Analysis and Planning

Microcomputers have changed population sector assistance efforts in two ways. First, microcomputer portability enables project staff to undertake analyses, revise population development models, and present findings as part of their short-term, in-country assistance work. Several projects help developing countries to assess current demographic conditions, project future demographic conditions, and evaluate population and development relationships. This assistance includes collection and tabulation of demographic data, analysis of population dynamics and population development relationships, dissemination of analysis findings, and incorporation of population considerations into development planning work.

Under the recently completed World Fertility Survey and Contraceptive Prevalence Survey projects, the microcomputer is being used to analyze data. Westinghouse Health Systems, under AID's Family Health and Demographic Surveys project, is assisting in the collection and analysis of demographic data and using microcomputers for entering, editing, tabulating, and analyzing survey data (Radloff, 1985). Where commercially available software is deemed insufficient, this project is also developing or adapting software to perform these data processing functions.

Also, Westinghouse Health Systems, under AID's Demographic Data for Development project, helps statistical offices make better use of existing demographic data. Westinghouse has transferred microcomputers and provided training in their use for analyzing data when statistical offices lacked computer time or training. In addition, Westinghouse has mainframe software programs for estimating demographic parameters and for performing population projections. The project has made these programs available to statistical offices, provided training in their use, and extended technical assistance in analyses using these programs.

Under the same project, the International Statistical Programs Center of the United States Bureau of the Census provides assistance to developing countries in conducting national population censuses. This project has been oriented towards mainframe computer equipment and has included development of mainframe editing and tabulation software programs. Recent advances in microcomputer capabilities for handling large data sets make it increasingly feasible to assign certain subsetted census tasks and even, in smaller countries, to process the entire census on microcomputers.

The Futures Group, under AID's Resources for Awareness of Population Impacts on Development (RAPID II) project, provides technical assistance to developing country policymakers in evaluating population trends and development implications. The Futures Group has used the microcomputer for preparing and displaying graphic representations of population and development relationships. Graphic display programs permit these implications to be demonstrated to policymakers.

Statistical Software for Demographic Analysis

Although there is general agreement about the core of techniques to be used for demographic analysis, a consensus on specific programs is difficult to reach. A review of required software was carried out by Michael A. Stone of the Population Studies Center of the University of Pennsylvania (1984). The respondents suggested that a statistical package, usually SAS or SPSS, would be necessary to perform their work. They agreed on a program, or set of programs, implementing the following statistical procedures:

- Descriptive statistics on individual variables, including mean, standard deviation, median, and mode
- One-way frequency distributions, with optional descriptive statistics and histograms
- Cross-tabulation (contingency) tables of at least three dimensions
- Correlation (and perhaps covariance) matrixes
- Multiple regression
- Graphic presentation of results and data.

One aspect of the statistical packages that virtually all of the demographers surveyed found necessary were functions to transform or recode variables, especially dummy variables, and select certain cases for analysis. Other useful data-related capabilities were the ability to save these transformed or created variables, to use parameters produced in one procedure as data in another, and to work with the residuals after regression procedures. Finally, almost all of the statistical procedures should be able to save their results in a data file for use by other programs.

Controlling Duplication of Data

In statistical analysis, reliability of data is crucial. Before any data set can be analyzed, duplicate records must be eliminated. A group at the faculty of engineering, University of Peradenlya, Sri Lanka, studied the feasibility of using a microcomputer program to remove duplicate entries from a set of clinical records (Gunawardena et al., 1984).

Based on a study of a sample of clinic records from the General Hospital, Kandy, Sri Lanka, in the early 1970s, it became clear that a systematic search for duplicate data would have to be carried out. Different methods of comparing the records were studied by simulation, using the actual duplicates discovered in testing to evaluate the effectiveness of various methods. The simulations were originally conducted using an IBM 1130 mainframe with 32K of core memory and 0.5M of disk storage. Today, however, those capacity requirements can be exceeded by a microcomputer, and the program is being run on an Apple IIe. The use of an interactive system makes it possible to detect a potential multiple entry in a few minutes, so clinic personnel can search the files as part of the patient sign-in procedure.

Data obtained from clinic patients is compared with records already entered. Previous entries are linked by comparing the following information in order of importance: sex, ethnic group, last name, initials, and age. Records are written in the Latin alphabet, so a subprogram has been designed to regularize spelling when transliterating from other alphabets. This program involves a number of steps including omitting all vowels and treating consonants that are used interchangably, such as "G" and "K," as identical. The newly entered record is compared with the records in the master file, and if no match is found, it is entered as a new file. If a match is found, the number of the earlier record is entered in a box on the new record form, linking the two records.

Modelling and Simulation

The Need For Modelling

Modelling, or the art of describing physical or other systems with equations, usually for the purpose of planning and decision making, is one of the most complex uses of computers, both with respect to the computations and the reasoning behind them. One broad class of model is for systems simulation, that is for predicting the behavior or outcome of a system based on its characteristics, some of which may be controllable. Examples include the following:

- Models for predicting blood pressure as a function of age, height, and weight
- Health as a function of quantity of water used for hygienic purposes
- Birth rate as a function of economic status and education
- Pressure in piped water networks based on flows, pipe diameters and ground topography.

Another broad class of models assists in the following functions:

- Allocation of nurses to duty stations in a hospital
- Allocation of a budget among alternative expenditures
- Determination of pipe sizes in a water network to minimize cost.

Developing and solving these models involves complex computations for which the computer is almost indispensable.

Once a model has been developed, it can be used for simulation by inserting hypothetical data and calculating system outputs. Since the computer solves model equations so quickly, it is easy to determine the sensitivity of the system by looking at the effect on the whole of changing one or a few variables. Historically, models and simulations have been run on mainframes. However, since they allow for decentralization of research, microcomputers offer great advantages over mainframe computers because they provide results quickly. In the past, storage capacity limited their use, but as microcomputers are made more powerful, they are becoming capable of solving all but the largest models.

Many types of models have been developed in the health planning field. One type that is frequently set up using electronic spread sheets is a model for scheduling, for example, the assignment of nurses in a hospital to ensure a sufficient number with proper skills in each ward on each shift. Scheduling is an optimization problem where a solution is sought using a mathematical expression with a series of constraints such as money, resources, physical quantities, and policy restrictions.

The Research Triangle Institute (RTI) has made wide use of the microcomputer for other types of modelling. For instance, a human resources planning model was developed to help the Mauritanian government plan for education, health, food, population, employment, manpower, and the national economy. The model was initially designed for a mainframe but then adapted to a microcomputer. In Tanzania, a regional planning model was developed for the Arusha region, and another was prepared for Tunisia. These models use a microcomputer to help planners keep track of and make forecasts for education, health, water, population, agriculture, nutrition, land use, livestock, and forestry. (A database management system was also developed to organize the data necessary for the planning model.) RTI also has developed a number of microcomputer models that provide forecasts useful for development planning based on projections of population changes (Chao & Allan, 1984). The introduction of modelling into the planning process of cooperating countries is useful for achieving the goals of the Integrated Population and Development Planning (INPLAN) project.

A similar type of model is one described by Teel and Ragade (1983) for dynamic social systems simulation and population management in Bangladesh. This model is being converted to run on the microcomputer. Already, the Bangladesh Planning Commission is using the results of this model in their five-year national plan, and the World Bank is interested in using the results in their five-year national plan and for the work of their family planning program in Bangladesh. In addition to long-range planning, microcomputer models may also be used to assist in decision making.

Harold Sox, at Stanford University, is involved in developing a microcomputer decision-support system (Sox, 1984). Most of the Sox research has been in the area of clinical algorithms (step-by-step instructions for solving a clinical problem), which lend themselves well to microcomputers. Parallel to this are "expert systems," applying artificial intelligence methods to medicine for diagnostic purposes. These systems have been developed on large computers at the University of Pittsburgh (INTERNIST), and at Stanford University (MYCIN and ONOCIN). These programs use a set of rules to interpret clinical data instead of following a specified sequence as in the case by clinical algorithms. These programs have not yet been adapted to microcomputers, but as microcomputer technology becomes more advanced, there is every reason to expect they will be.

Environmental engineers use models for such things as the design of water distribution networks, water quality studies, river basin planning, and other aspects of water resources work that have an impact on health. Both simulation and optimization models are common; the objective function for the latter is usually an expression of cost to be minimized.

Computer modelling and simulation can also teach medical students, engineers planning water systems, government planners, and economists very effectively. By examining different scenarios, planners can forecast the consequences of possible decisions, in effect accumulating surrogate experience rapidly.

Population Modelling

Most socioeconomic-demographic models are a series of mathematical relations that usually represent population growth rates as a function of several variables (such as birthrate and health care). When put on a computer, models can aid planners and decision makers to address development problems facing the country. They are equally useful to the planner in producing a consistent set of numbers in a given framework and in assessing the implications of alternative policies. Moreover, models provide planners with the means to project future events that avoids the tedium of a complicated series of calculations that might not be made, or would only be made rarely, if done by hand. For the policymaker, models can illustrate the importance of changes in fertility, mortality, and internal migration to the development of the economy.

RTI, under the INPLAN project, provides assistance to developing country planning institutions for understanding population and development relationships and for incorporating population considerations into development planning work. These models include a multiregional population projection model, a migration sector model, an education model, a manpower and employment model, a health services model, an agriculture model, and a family planning and health status model.

Under the first phase of the INPLAN project, RTI developed a multisectoral microcomputer model to evaluate the costs and benefits of a national family planning program. The purpose of this model is to examine the effectiveness of a country's family planning program in reducing the costs of social welfare spending while improving the quality of the social services provided. Although policymakers may recognize the need for a strong national family planning program, they still may hesitate to allocate government resources from other sectors because the outputs from the program are not tangible. If it can be demonstrated that a strong family planning program could generate substantial savings in government expenditure, policymakers and planners in other ministries are more likely to become financially committed to a strong program.

The model uses both commercial software (LOTUS 1-2-3 spreadsheet modelling) and project-developed software, including:

- Multi-Regional Population Projection Model: projection of population growth by region
- Human Resources Planning Model: estimation of future needs and resource requirements in education, employment, nutrition, and health
- Education Planning Model: estimation of future education needs and required resources
- Labor Force and Employment Planning Model: estimation of future employment needs and required resources
- Health and Family Planning Services Model: estimation of future health and family planning needs and required resources
- Cost-Benefit Model: estimation of cost-benefit ratio of family planning delivery programs.

The AID-sponsored models, Resources for Awareness of Population Impacts on Development (RAPID-II), was developed to present information in decision-supporting models to developing country policymakers. The models provide technical assistance to developing country policymakers in evaluating population trends and development implications. The Futures Group has made use of the microcomputer for preparing and displaying graphic representations of population and development relationships for awareness-raising and dissemination purposes. These graphic presentations have been used for illustrating current and demographic characteristics of a population; implications of demographic change for specific development sectors; and contraceptive prevalence rates and required increase in prevalence to achieve future fertility target levels. Interactively designed presentations permit on-the-spot evaluation of the effect of altering underlying assumptions of demographic and economic parameters in relationships.

RAPID-II also uses LOTUS 1-2-3 for spreadsheet modelling and graphics display. Project-developed software includes:

- Rapid Presentation Model: estimation and display of socioeconomic relationships
- Demographic Project Model: projection of population growth
- Fertility Target Model: estimation of required contraceptive use to achieve a future fertility level
- Education and Demographic Simulation Model: estimation of future education needs
- Food Model: estimation of future food production needs
- Cost-Benefit/Cost-Effectiveness Model: estimation of cost-effectiveness and cost-benefit ratio of family planning delivery programs.

Models for Research

Computer modelling can assist a researcher in studying the effects of varying inputs to a system. For example, at the Malaria Research Center in Delhi, a model has been constructed to incorporate the effect of seasonal fluctuations on the rate of change of malaria vector populations (Srivastava and Sharma, 1984). A parameter called "force of infection" incorporates the level of incidence of malaria, a function of the relative density of the vector and host populations. The model provides two scenarios: prevalence of mosquitoes in the absence of any intervention, and the effect of spraying of a residual insecticide. The model also reflects the fact that the effect of spraying diminishes over a known period of time. This model can be used to predict the severity of malaria incidence at a future time when the existing mosquito population is known, or to study the effects of spraying or other interventions made at various points in the cycle or at varying intervals of time. Future refinement of the model will allow the researchers to perform rapid assessments of other proposed interventions and to demonstrate cost-benefit ratios for their use.

A more complex research model was designed by Fernando and Sandra C. Bertolli (1981). The model, based on a statistical analysis program designed for the Ministry of Health in Morocco, demonstrates the causes and prevalence of four water-borne diseases: malaria, typhoid, dysentery, and schistosomiasis. They used the Interactive Statistical Inquiry System (ISIS) on an Apple II

microcomputer to look for an explicit relationship between incidence of the four diseases and socioeconomic factors. The study
was designed to aid planning by identifying the social constraints
that affect health levels.

For each disease, reports from each province over a three-
year period were averaged and divided by the population of that
province. The model enabled researchers to study such questions
as the actual incidence of the four diseases in each province, the
interrelationship of disease incidence in various areas over time,
and the correlation of predictors of the four diseases. One early
outcome showed that provinces with a higher number of households
with running water had lower incidences of the diseases under
study. This factor was shown to be more important even than the
presence of doctors.

These first returns from the model point out the need for
more accurate and complete data collection. As more uses are
discovered for the model, more diseases can be added to the database. Finally, when larger amounts of data and more varied data
are collected, there may well be a need to analyze them at a
subprovincial level to avoid long delays and permit local verification of accuracy.

Models for Environmental Systems Design

The use of mainframe computers in developing countries is
hampered by their high cost and complex environmental requirements. When systems design models became available on microcomputers, however, they became accessible to a great many
potential users. Government agencies and engineers soon find that
microcomputers can be used with a minimum of training and can be
applied to many technical problems in the planning, design, and
operation of systems. The accessibility of microcomputers made it
possible for design engineers to study a number of alternatives to
determine the most efficient or lowest cost approach.

Modelling is proving of great value in the design of water
supply systems (Hebert, 1984). Developing countries face a range
of problems in trying to meet the goal of the United Nations
Drinking Water and Sanitation Decade (1981-1990)--to provide
adequate water and sanitation services for all by 1990. There is a
need to maximize the population served by making the best use of
existing budgets and to maximize the productivity of scarce technical experts in determining appropriate service standards and
design criteria. A major constraint is that services must be provided at minimum cost so they are affordable by low-income
users. Engineers in many developing countries currently design
complex systems using hand calculations and have no means of

quickly and accurately estimating costs. These tasks are time consuming, making it difficult to consider alternatives. The consequence, therefore, may well be a design that is overly conservative and more costly than necessary.

Over the past several years the World Bank and the United Nations Development Programme (UNDP) have been actively promoting the use of computers by water supply agencies in developing countries to improve planning and design. The assistance has included the development of computer programs, training of agency staff in their use, assistance with purchase of microcomputers and with routine application of the programs. Developed at the University of North Carolina at Chapel Hill, the programs have been tested by water agencies in the Philippines, Indonesia, Thailand, Burma, India, and Sri Lanka. The following is a description of these technical assistance and training projects and the lessons learned from them.

Water Distribution Systems

The World Bank/UNDP Project has focused on assisting agencies to improve water distribution network design. Design is one of the most technically difficult phases in project preparation, and since distribution systems often account for more than one-half of the cost of an entire water supply system, efficient design is important.

Distribution networks in urban areas are usually looped because pipes are placed in most streets adjacent to houses where connections are made. There are numerous methods for designing looped distribution networks. One of the most widely used is the Hardy-Cross method, which, through an iterative procedure, makes adjustments in pipeline flows until the friction losses around each loop in the network sum to zero and the inflow and outflow for every node in the network are balanced. By hand, the method is extremely time consuming, and, therefore, alternative designs cannot be easily investigated.

The World Bank/UNDP Project developed a network design/analysis program, called LOOP, suitable for microcomputers. The program, written in BASIC, can be used by engineers with no previous background in programming or computer use. The user specifies the network geometry, the lengths and trial sizes of pipelines, the demands, and other basic design parameters. The program then generates the hydraulic characteristics of the network, including pipeline flows, friction losses, velocities, and pressure at each node in the network. The user can test alternative pipe sizes and/or pump characteristics until a satisfactory flow and pressure distribution is achieved.

Design time can be cut to a fraction of that required for hand calculation. More importantly, numerous alternatives can be evaluated; these might include alternative locations for storage tanks, sources, and pipeline staging to arrive at a least-cost design. The major drawback to the program is that several submissions to the computer are necessary, and an optimal design is not guaranteed. Other programs written in BASIC by various private parties are also available for network design.

In rural areas, where population densities are lower and where service may be from public faucets rather than by house connections, the distribution network is usually laid out in a branching or tree configuration. For branching networks, the flows in pipelines are known once demands at the nodes in the network are defined. An optimization technique can be used to determine pipeline sizes that minimize network costs while satisfying specified minimum pressure and headloss contraints. Linear programming has been used successfully for this purpose. The World Bank/ UNDP Project has developed computer programs, written in FORTRAN and in BASIC, for both mainframe and microcomputers for optimal design of branching networks, using a linear programming algorithm. The program, called BRANCH, has been applied to the design of rural water supply schemes, bulk transmission schemes, and for designing trunk mains of larger urban water supply projects in several countries of South and East Asia. The program selects from among specified commercially available pipe diameters and determines the least-cost combination of lengths of each size pipe in each line to maintain the required minimum pressure at terminal nodes in the network.

Cost and Financial Analysis

Economic and financial analyses of projects are time consuming tasks if performed by hand, and only limited sensitivity analyses are possible without the aid of a computer. Electronic spreadsheets, such as Visi-Calc, SuperCalc, Multi-Plan, and Lotus 1-2-3, are available for performing cost and financial analyses with microcomputers. The UNDP/World Bank Project has assisted agencies in the Philippines and Thailand in developing custom spreadsheets for financial and cost analysis. Typical financial statements that have been developed include Income Statements, Cash Flow Statements, and Balance Sheets. The project has also developed a simple multi-linear regression analysis program, which has been used to develop cost functions for various water supply components.

Other Technical Applications

The UNDP/World Bank project has also developed sewer network design programs written in BASIC, suitable for both conventional and small bore sewers. Engineers in the agencies that have received assistance are also developing their own programs, primarily in BASIC, for such problems as the structural design of storage tanks, treatment works, and other civil works and for the analysis of groundwater potential.

In 1982, because of problems in gaining access to mainframe computers, it was decided to shift emphasis to the use of microcomputers. Several machines were purchased and programs were quickly converted to run in BASIC. Although some processing speed was lost when compared with the large mainframe computer, job turnaround time (the time interval between submission and receiving printed output), decreased by a factor of 10. The network design programs now have been modified to run on numerous microcomputers, including the HP-85, HP-86/87, Osborne 1, Cromemco II, DEC VT-180, DEC Rainbow, Apple II, IBM-PC and all IBM-PC compatible machines, and the Wang PC. Conversion to run on other computers will be quite easy because the programs are written in BASIC, and although BASIC may differ from machine to machine, the differences are usually minor.

The World Bank/UNDP project has provided technical assistance to the Philippines, Thailand, Indonesia, Sri Lanka, Burma, India, and the People's Republic of China. In addition to modifying the LOOP and BRANCH programs for local circumstances, the project offered help in selecting hardware and in training staff.

Agencies in these countries have noted numerous benefits, including a reduction in the time required to prepare project plans and detailed designs, more efficiently designed and less costly distribution networks, and more efficient use of engineers and technical staff.

The experience of this project suggests that several steps are necessary to ensure that microcomputers are used effectively in water sector organizations. Initial demonstration of programs relevant to the engineering and other operations of the organization is necessary, followed by training of staff in the use of the programs. Assistance is also necessary to help identify appropriate computers and peripherals and acquire funds for procurement. Finally, help may be required to modify programs for computers that are purchased.

Thus far, the project has dealt primarily with the technical applications for microcomputers in the water sector. Clearly, there are other areas where microcomputer applications could result in significant benefits. Monitoring of project preparation,

construction, operation and maintenance, and financial performance are related areas where simple database management programs could greatly increase efficiency in most water supply
agencies. Billing and collection is another area where computerization can be beneficial.

Risk Assessment

For health resources to achieve maximum benefits, they must
be properly allocated. One approach is to provide health services
first to those who are at highest risk (Daly, 1984). Therefore,
those patients who would otherwise have the highest risk of low
birth weight and infant mortality can be given preference in prenatal and well-child care. Similarly, secondary care can be
allocated according to risk, as in the case of hospitalizing only
women with high probability of complications of delivery or of
delivering an infant of low birth weight.

Increasingly, paramedicals, often with relatively little formal
education or medical training, provide primary care, and are
required to make decisions on the referral of patients for more
intensive outpatient or inpatient services. With the advent of
low-cost, reliable, hand-held microcomputers with significant
memory capacity, it becomes possible for paramedicals to recall
and organize the necessary data. Moreover, microcomputers
greatly facilitate the processing of epidemiological studies needed
to derive quantitative risk indicators.

Risk Indicators

The Office of the Science Advisor in AID, through the National Academy of Sciences, is funding a number of studies to
develop risk indicators. Osmand Galal of the Egyptian Nutrition
Institute is studying risk indicators for fetal malnutrition; Edgar
Kestler of the Guatemalan Social Security Institute is developing
an instrument to identify women's risk of delivering low birth-
weight infants; and Francisco Mardones of the University of Chile
is studying instruments to predict malnutrition in the first year of
life.

The development of a risk indicator depends on an epidemiological study in which a population is described by a number of
"predictor" variables and an "outcome" variable. In theory, information on the predictor variables should be available to the field
worker actually evaluating risk. The outcome variable is that
condition for which the risk is to be estimated. When predicting

risk of future malnutrition, for example, one may consider predictor variables measuring current nutritional status, ability to acquire food, diseases influencing nutritional status, and nutritional requirements. Other variables, perhaps more easily measured but more indirectly related to malnutrition, include indicators of family financial, social, educational, and other capacities, and environmental and hygienic variables. In practice, the selection of the specific set of indicator variables for the risk indicator is perhaps best accomplished sequentially in the definition of the indicator itself.

Case-control studies have proved the most efficient in providing the data for estimating risk. In these studies, a random sample of people suffering from the condition to be predicted is identified. A sample of other individuals from the same target population is also selected. Matched by variables such as geographic residence, age, and sex, the potential predictor variable values are evaluated for both cases and controls. The study population is then divided into two random samples, one to develop the risk indicator and one to test its accuracy.

Several alternative methods have already been described for the development of risk indicators on the basis of this data (Fisher, 1936; Feldstein, 1966), and a relatively rapid procedure has been adapted from a pattern recognition algorithm (Nelson et al., 1965).

The application of these simple risk approaches to health service delivery systems has limited potential but does appear to be worthwhile for relatively unskilled paraprofessionals. Thus, an outreach worker trained to deal with prenatal visits, well-baby visits, nutritional rehabilitation visits, oral-rehydration therapy, and vitamin-A supplementation might be provided with a risk index for each class of patient (to help in the decisions as to which patients to leave alone, which to treat in the field, and which to refer for further diagnosis and therapy to professional practitioners).

Ancillary Functions

In addition to its use as a stand-alone machine, the microcomputer can be used in conjunction with other equipment. For example, microcomputers are often used in hospital monitoring to transform and display the signals received from measuring devices. They are widely used with CAT scanners (computed axial tomography, a special X-ray procedure); microcomputers are also being used in conjunction with electrocardiograms, measurement of respiratory signals, and more. Computer analysis of phonopulmograms apparently may prove to be a promising noninvasive diagnostic tool for lung diseases (Majumder, 1980).

83

Another example of computer use to process information from monitoring equipment comes from the People's Republic of China where research was begun in the early 1970's on acupuncture anesthesia. Computers were used to transform and analyze the waveforms of electro-encephalographic signals in order to relate certain wave patterns to specific types of acupuncture. The work could now be done on microcomputers instead of mainframes.

The microcomputer has also been used to control videocassettes for health education purposes. In the United States for example, the National Hansen's Disease Center in Louisiana produces educational programs designed to teach early recognition of Hansen's Disease (leprosy) and various aspects of patient care. Recently, they have been producing videocassettes that have been mailed to several countries and translated into several languages. Although these programs are meant primarily for use in centers, seminars, or schools, the Center has recently been experimenting with microcomputers to control the videocassettes and change the program into interactive instruction for individuals. Rather than present the entire film in given order and at set speed, the micro-computer breaks the film at various points to ask questions; depending on the viewer's response, the machine branches to dif-ferent segments of the film to reinforce or correct the viewer's answers. This combination of learner involvement through the microcomputer's interactive process has made this technique quite successful in the trials that have been conducted at the Louisiana Center.

Expert Systems

Both in the effort to provide useful tools for physicians and to develop a science of cognition, researchers in artificial intelligence have developed a number of "expert systems" for medical diagnosis and prescription (Barr and Feigenbaum, 1982). These computer programs normally include a knowledge base, incorporated in a number of rules, and an analytic procedure. For example CASNET, developed by researchers at Rutgers University for treatment of glaucoma, includes information on about 150 disease states and 350 signs, symptoms, and tests of glaucoma. It links symptoms and test results to allow probabilistic judgments to be made about the con-dition of a glaucoma patient and the severity of the condition to be tracked.

The development of expert systems for medical decisions typically involves the participation of highly qualified, board-certified medical specialists who also have interests in clinical research in computer science. It also involves extensive periods of testing with real patient records or in some cases with real clinical situations. The programs also have extensive capabilities to acquire new data, interact with users, explain their logic, and adapt their structure to accommodate new information.

It seems likely, however, that a fully developed system could be greatly simplified to adapt it for use by large numbers of people. In terms of memory requirements and machine speed, it would appear that expert systems for limited aspects of medical decision making could easily be adapted to microcomputers, and even to hand-held microcomputers that could be used in the field.

To date, very few of these systems have been used in clinical medicine. In part, this is a result of the novelty of the research, which is only about 10 years old, and in part, it stems from the fact that specialized physicians in the United States and other developed countries have not felt the need for the existing systems to carry out their professional duties. It has been suggested, however, that the first major applications of expert systems might well occur in developing countries, where the lack of trained medical specialists makes this "consulting service" even more valuable.

One country that is vigorously pursuing the idea of diagnosis by microcomputers is the People's Republic of China. China is well known for its traditional doctors. Their numbers are declining, however, and there is now an effort to record their accumulated body of knowledge before it is lost. Shires (1982) reported after his visits to the Chinese Medical Informatics Association that the first programs were written for hepatitis, cardiovascular disease, thromboembolic disease, arthritis, and skin diseases. By 1982, roughly 40 clinical algorithms existed, simulating human decision making for diagnostic purposes. According to Shires, the Chinese are planning to develop about 50 of these programs for use on small computers, to field-test and validate them and to distribute them to small community hospitals across the country to ensure a uniform level of patient care.

Future Directions

A necessary step in implementing expert systems or computer-aided diagnosis is to define norms and standards for the allocation of medical resources that maximize the benefits derived by society. Ideally, then, one would seek a set of treatment norms within the available resources, so that any change would result in a net loss of patient welfare. Taking resources from one group and allocating them to another might result in more loss to the first group than gain to the second. Thus, groups would be defined not only by their risk of harm without treatment, but by the relative risk to the population as a whole with different allocations of resources. In fact, physicians are willing and experienced in making such judgments. The patient who informs the doctor he cannot afford the normal treatment may rationally ask what an alternative, less expensive treatment might be, and request to know both the cost and the prognosis under both alternatives. In designing systems to simulate this decision making process the ability of microcomputers to embody knowledge and analytic capacity and the ability of minimally trained people to carry out expert functions using these micros should greatly contribute to the quality of medical care in developing countries. However, it should be emphasized that expert systems require local adaptation to fit them to local cultural norms and perceptions about disease.

SELECTED EXAMPLES

Rural Health Delivery, Central and West Africa

One of many organizations that rely on microcomputers is the Project for Strengthening Health Delivery Systems (SHDS) in Central and West Africa. Funded by AID, SHDS was developed at Boston University to respond to the need expressed by 20 Central and West African countries to improve regional and national disease surveillance, health and demographic data systems, and to integrate these systems into national health planning systems (see Figure 4-2). SHDS has found microcomputers to be useful in its daily operations and is encouraging their use in project offices (French, 1984).

In the project offices, microcomputers are used to track ongoing budgetary and fiscal activity. Visi-Calc and D-B Master software on the Apple II microcomputer have been adapted to follow the detailed implementation plan and project budget formats in use. The data entry is done entirely by the African support staff, adding a new dimension to their office responsibilities.

Microcomputer programs for various substantive health information systems have been developed for different countries, including a Field Epidemiology Training Program and a program for data collection from peripheral health centers in the Ivory Coast and a model health information system program in Sierra Leone.

The most significant aspect of this project may well be not the technical information derived from it but the transfer of technology to the public health sector, which is, by its nature, deeply involved with local realities. In the regional SHDS office, the objective of training of the African office support staff was the transfer of the ability to: (1) configure software for specific applications; (2) determine whether operational problems were user, software, or machine related; and (3) master the manual of instructions so as to minimize recourse to external sources for solving these problems (Helfenbein, 1984).

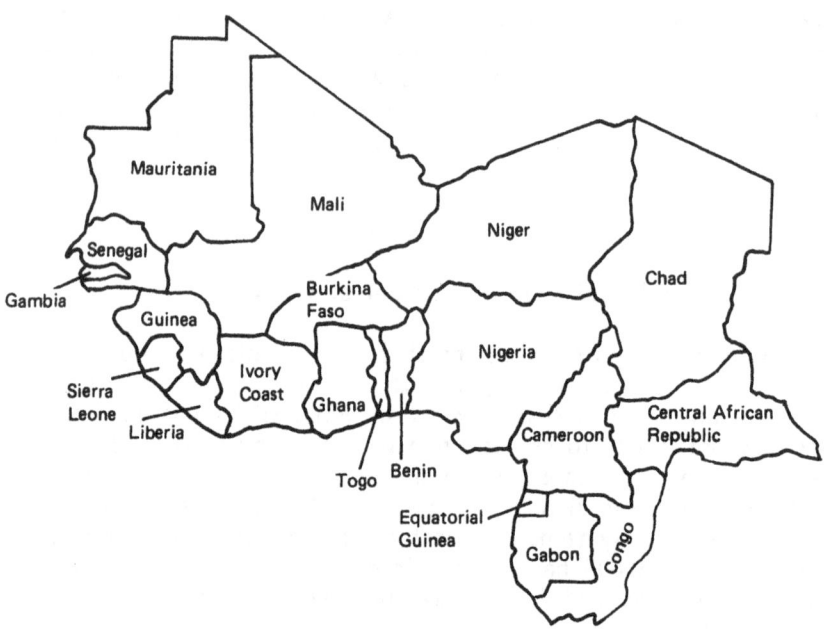

Figure 4-2   Countries participating in strengthening health delivery systems (SHDS). (Newsline, June 4, 1982)

Training, it was discovered, must be considered in three time periods. In the short term, it should aim to develop skills to operate a desired application of a given software. In the medium term, the user should master the given software so as to be able to configure several applications. Finally, in the long term, the user should develop diagnostic capability and be able to determine user-, program-, and machine-related problems. While the training must be supplied initially by an expert, continued cooperative training or user-user interaction proved to be very effective, since it tends to reinforce recently acquired skills and develops a team approach to problems.

The experience of this successful small office training project suggested that essential features for success include the following factors:

- Defining the decisions that have to be made and the information that needs to be collected
- Encouraging users to teach themselves by studying the manuals so that they can move beyond the initial applications to a mastery of the system
- Identifying one or more users who are willing to train co-workers and help them when problems arise; and
- Creating an environment in which the use and ultimate mastery of this technology is encouraged and rewarded (French, 1984).

District Health Data Analysis, Malaysia

Soon-Teong Teoh of the Faculty of Medicine of the University of Malaya in Kuala Lumpur, Malaysia, is part of a group involved in setting up a microcomputer-based system for the reporting and analysis of health data and statistics. Currently, various data, such as costs, personnel, births, illnesses treated, lengths of hospital stays, and bed occupancy, are collected at the district level. These data are "sent up" to provincial and national levels for analysis, and reports are "sent down" to the district levels to assist with planning and decision making. In fact, there are long delays in obtaining reports from the central agencies. Consequently, district agencies have obtained their own microcomputers, and with help from Dr. Teoh and his colleagues, they prepare their own reports and analyses in order to obtain information in a timely manner (Teoh, 1984).

The use of microcomputers can help to establish the flow of information between centralized health facilities in the capital or at the university and field clinics. However, the usefulness of such an information exchange will depend on the type and accuracy of field data, its processing, and the type of information returned to the field. The danger exists that field data may be used only in the capital to make health care policy and that no reports will be returned to individual health centers. This project was begun to counteract the lack of feedback from data submitted by rural health clinics.

The University Faculty of Medicine conducts a number of undergraduate as well as postgraduate training programs in a health district some 50 miles from the campus (see Figure 4-3). This district--the Kuala Langat Health District (KLHD)--has a full complement of health facilities, (rural clinics, health centers, a health office, and a district hospital). It is well within daily traveling distance from the campus. An examination of the current routine system showed that what was suspected was indeed true. The district health services collected much data and transmitted it, but received little feedback in return. It was then

decided that to examine all health data would be too big a task, so
the project began with the collection of maternal and child health
data.

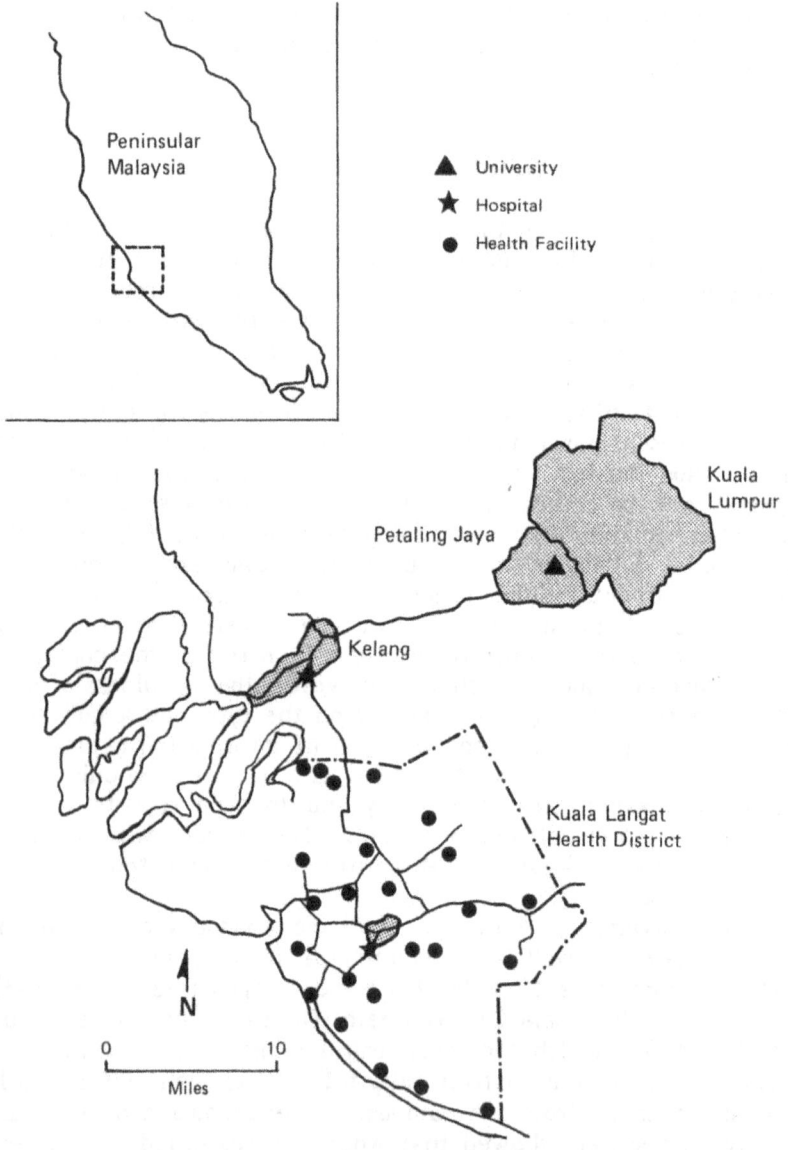

Figure 4-3   Study area location

KLHD contains a total of 23 health facilities, all of which conduct programs that cater to the health of mothers and children and collect relevant data. The traditional system employed in recording the data is depicted in Figure 4-4. It can be seen that at each stage of the system written records were kept, starting with counts recorded in exercise books at the rural clinics. The clinics send the reports to the health centers and the information is recopied into ledgers. The health centers then send these ledgers to the health office where the same records are re-entered by hand into a master ledger.

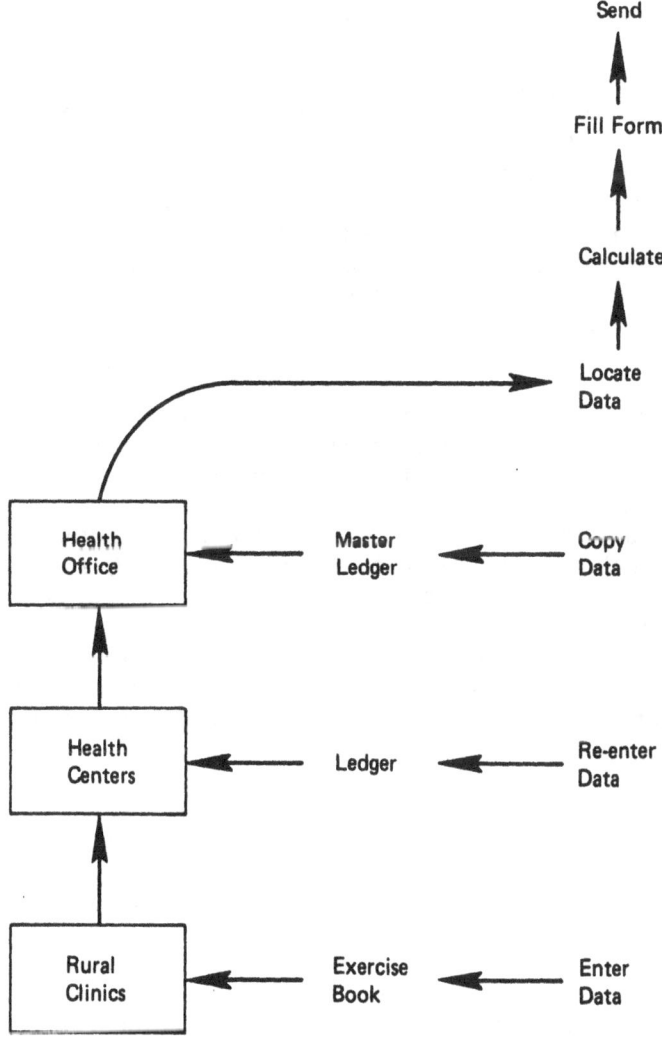

Figure 4-4   Current health data system

Each year, at various times (monthly, quarterly, semi-annually, or annually), a total of 16 different forms is sent to and filled out by the maternal and child health service staff at the district level; the forms are then sent to the Ministry of Health. It became obvious to the University of Malaya team that the traditional system of processing data concerning maternal and child health is time consuming, laborious, and prone to clerical and arithmetic errors. For these reasons, a system was designed to make data processing easier and faster. It was decided to use the data from the maternal and child health services to run a pilot project test of the new system.

The database contained the following items:

- Health Data—pertaining to the activities of the health services in relation to the community and to patients who attend the various clinics and the district hospital.
- Administrative Data—dealing with the health staff employed, vacancies, payrolls, leave of absence, inventories, financial statements and the like.
- Background Data—such as population by age, sex, ethnic group, occupation, geographical location, and financial status.

It was obvious that the proposed system required a microcomputer for the data to be stored and retrieved quickly. The microcomputer also made it possible to generate the required indexes with which the KLHD could evaluate its own performance (see Figure 4-5).

The team then reviewed the various microcomputer systems available in the country at that time (January 1983) and decided that an APPLE II+ with two disk drives and a dot-matrix printer would be suitable. It fitted most performance criteria, and a number of similar APPLE-compatible machines were also available at a lower price. dBASE II was selected as the appropriate software. Three other smaller projects were subsumed under this project and utilized the microcomputer system:

- An epidemiological surveillance and epidemic data analysis program, which was adapted from a program from the Centers for Disease Control (CDC) Atlanta, Georgia, originally written for the Kaypro portable microcomputer
- A program for the analysis of some in-patient data from the Banting District Hospital, which serves the KLHD
- A program for collating data on vector control and malaria cases in the KLHD.

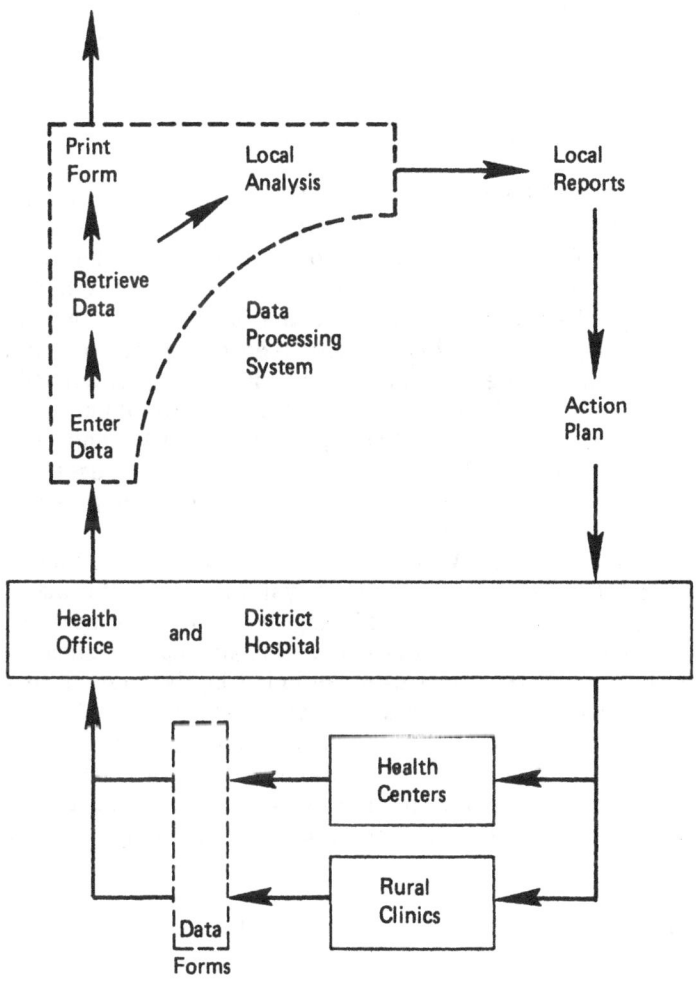

Figure 4-5   Proposed health data system

The first two were already in use by October 1984, and data from an epidemic of German measles, which occurred in the University campus in early 1984, was compiled and analyzed by the program adapted from CDC. This project, with its sole study area in the KLHD, thus far seems to have shown that there is a failure of the feedback loop in the routine data flow system from local health district to central government and back. The creation of a local data processing system to supplement (but not disrupt) the routine data flow system seems to be an acceptable solution.

Hospital Management, Thailand

The Ministry of Public Health (MOPH), Thailand, is developing a hospital information system to support more effective hospital administration. The use of health information for planning, monitoring, and evaluation of health development projects has been recognized by Thai health administrators since 1977, when the World Health Organization (WHO) introduced the system of country health programming. Before this period, data were collected and compiled, but without being used either in health planning or in decision making. Under the WHO program, health and health-related data were systematically collected, processed, and analyzed in an effort to develop a sound health development plan for one province (Boonthai, 1984).

In addition, in 1983, MOPH decided to introduce computerization of its information management system. A master plan was formulated to cover all data, including hospital information, requested for electronic data processing. Microcomputer subsystems for provincial and district use were included in the master plan in order to link the central computer system to provincial centers. As pioneers in this new highly technical development, some hospitals within the jurisdiction of the Ministry of Public Health have brought in microcomputers at their own expense to support hospital management. These systems are used for keeping records of various kinds:

- Medical information for both inpatients and outpatients
- Personnel management
- Control of inventory and the stock of drugs
- Budget control and accounting
- Research work

The use of microcomputers in those hospitals has been initiated individually, so the systems are not standardized. This highlights the need for coordination in hospital management. In Thailand, in an effort to respond to problems caused by the lack of standards, a National Computer Committee was established to authorize all government sector requests for microcomputers and to monitor and control all official computer systems. The Ministry of Public Health has formulated a master plan for computerization of data processing services in all health units in the ministry, and has arranged to conduct training courses for health personnel.

The National Computer Committee conducted a survey to identify the types of information that must be included in a computerized system. The results were as follows:

- Basic identification information for each patient
- Basic information on number of patient beds occupied and vacant
- Basic financial and accounting information including inpatient's expenses, budget allowance, hospital income, and budget expense by categories
- Basic information on drug stock, including lists of drugs available, amounts of each drug prescribed, replacement stock ordered, prices for each item, and expiration dates of drugs in stock
- Similar information for other expendable supplies.

In selecting and developing software, certain important considerations must be noted:

- The program must be workable on a microcomputer, yet be expandable to fill future demands.
- The program should be able to read any data file using American Standard Code for Information Interchange (ASCII) and be able to build a data file using ASCII characters.
- Managers of computer systems or programmers should keep up with changing software technology in order to upgrade software as appropriate.

At present, software used in hospital management is often independently developed by doctors for their own specific purposes. Some categories of management are widely used, such as stock control of drugs and medicines and number of patients with certain problems. Computer experts in the hospital management field can select and develop the most appropriate software for these common functions. A pilot project in two or three hospitals to clarify the design of the system should encourage standardization of both hardware and software to ensure maximum utilization of data (Boonthai, 1984).

Maternal and Child Health Care, Bangladesh

The International Centre for Diarrheal Disease Research, Bangladesh, (ICDDR,B) uses a database system specifically developed to avoid the pitfalls of incomplete or defective demographic data. In a pilot extension project with the Ministry of Health and Population Control, ICDDR,B is collecting data to show the effect of a Family Planning Health Services Project that delivers comprehensive family planning services to women in their homes and provides maternal and child health services in nearby clinics. To monitor a maternal and child health care project, data are needed on every household in the four rural subdistrict test areas (Phillips et al., 1984).

The Sample Registration System (SRS) developed by David Leon, a consultant to ICDDR,B, is designed to incorporate comprehensive continuous editing that produces error statements to be fed back to field workers within a few days of data collection. Moreover, comprehensive cross-linkage of data permits flexibility in the use of data for a variety of practical applications without long delays in processing and analysis.

The manual portion of the SRS consists of information recorded by government village workers in a Household Record Book (HRB). Every household is visited quarterly and the record book is updated with service information data on vital events. The computerized SRS is maintained by ICDDR,B workers on a sample of the HRB households. Analysis of the sample provides demographic data and comparative analysis of the accuracy and completeness of the HRB.

The SRS was designed to overcome the limitations of existing demographic data collection methods. Three problems were found to impede the utilization of this data for research:

1. Demographic data systems are often limited by their own dynamics. For example, independent variables that should not be omitted include the impact of family planning, the consequences of childhood morbidity or malnutrition, and the intervention variables such as immunizations and health care. A continuous, broad-based demographic record becomes a valuable tool in policy research because an existing database often means that when a question is raised, little extra information needs to be collected. Researchers can therefore respond in a timely manner to policy questions.

2. Vital events are often collected in batches by type of event rather than being linked to individuals at risk. This aggregate data can be used to report vital rates, but

studies on the determinants of mortality or survival, for example, require data linked to individuals.

3. Batch mode systems often do not provide for cross-checking of data entered. Since demographic events are logically interrelated, software must provide for logical checking, so that, for example, births must have a mother in the household, and deaths occur only to individuals known to be at risk in the household under observation. Microcomputers can be used not only to store data, but to link and check the logical consistency of events.

Setting Up the SRS. The first stage of setting up the SRS in Bangladesh involved enumerating approximately 325,000 households. Data on head of household, names of members, sex, date of birth, marital status, education, occupation, and family relationships were collected for each unit. Based on these lists, computer-generated Household History Forms (HHF) were prepared and bound into Household Record Books (HRB) for field workers. Workers were trained to locate the households, interview household members, and code the HRB for vital events. The information recorded on the HHFs is computerized and cross-checked for logical consistency. Data passing all tests are archived, and HHFs failing logical tests are printed with error messages for field correction. Currently, field correction activities are complete within a week of the end of a round of visits owing to a system of continuous processing of data from the time of collection. Error reports require reinterview, correction of the erroneous data, re-editing and linkage, and finally, when consistent, archiving. Figure 4-6 shows the process.

A single-step procedure to input data directly to a microcomputer is being developed. Models and data files are well within IBM-PC limitations. Programs require less than 128K of storage and the total database less than 8 megabytes if all available data are loaded simultaneously.

The HHF provides the framework for all survey research in the Extension Project. Special studies of village women, intrafamilial communications, economic status, and other issues are conducted with each round, linked to the SRS, and archived, so that previously collected SRS data are accessible for any cross-sectional survey.

Some early results of the SRS research are given below.

96

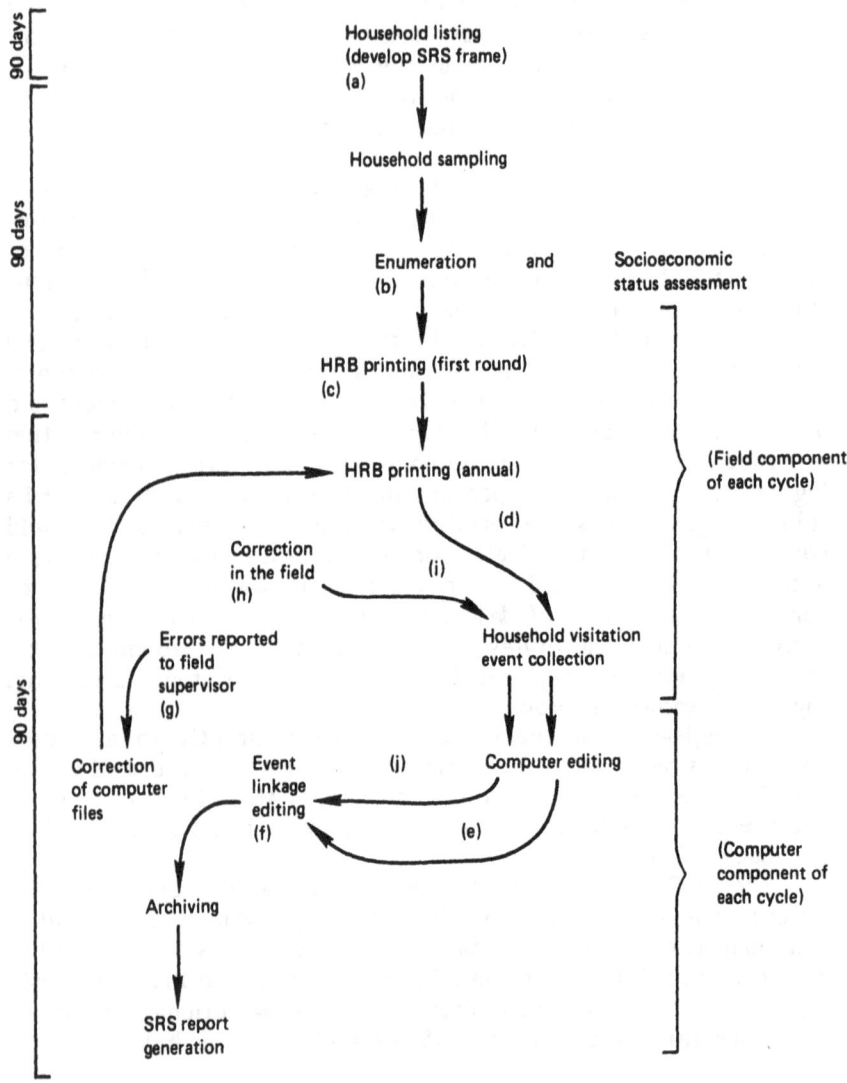

Figure 4-6   Tasks in SRS development

- **Mortality.** Infants born since 1982 were observed over the October 1982 to September 1984 period. A hazards regression study showed that sex of infant, economic status, and education of mother had no net effect on either neonatal or postneonatal mortality. Postneonatal mortality is high in households where siblings have died. Given the high total mortality in study areas, serial mortality in high-risk households has important policy implications. Further research is needed on the causes of serial mortality, possible interventions, and the probable impact of these strategies.
- **Fertility.** The Extension Project includes both high and low fertility areas. The socioeconomic determinants of this areal variation are the subject of the inquiry. In study areas, new approaches to family planning service delivery are being tested; among them is the use of female village workers for family planning service delivery. The net contribution of the introduction of this strategy to contraceptive use, and ultimately to fertility decline, is a subject of SRS analysis. For example, an early finding from the SRS is the noticeable importance of personal contact between clients and female family planning workers to contraceptive adoption.
- **Quality of Services.** Areas served by the Extension Project evince considerable variation in the quantity and quality of service delivery. The extent to which this variation is due to societal determinants versus relative operational deficiencies in the program is the subject of inquiry. Whether services can be improved by project interventions and the effect of changing service activities on health and family planning behavior is the subject of SRS-based research.

Environmental Engineering Design, Thailand

A program developed by Viswanath (1984) at the Asian Institute of Technology deals with the preselection of treatment processes for domestic wastewater and wastewater sludge. The alternative treatment processes include high as well as low technology options. The task comprises two main components: establishing feasible processes composed of successive subprocesses, and selecting the least cost solution (Orth, 1984). Table 4-1 lists the key parameters of the various treatment processes.

Table 4-1 Key Parameters of the Various Treatment Subprocesses

Process	No. of Loadings	Key Design Parameters	Cost Parameters
Primary sedimentation	3	Removal efficiency of SS	Surface area, flowrate
Activated sludge	9	Mixed liquor volatile SS Recirculation ratio	Flowrate, tank volume, air flow
Trickling filter	8	Hydraulic loading, Depth of media	Flowrate, volume of media
Rotating biological contractor	9	Hydraulic loading Number of modules in series	Surface area
Aerated lagoon	6	BOD loading Detention time	Flowrate, volume, Installed power
Anaerobic-facultative -maturation lagoons	16	BOD loading Detention time	Required land area
Facultative-maturation lagoons	6	BOD loading	Required land area
Gravity thickening	8	Solids loading, solids recovery Under flow solids	Surface area, dry weight of solids
Anaerobic digestion	4	Detention time	Volume, dry weight of solids
Vacuum filtration	8	Loading, cake solids, lime dose, FeCl dose	Filter area, dry weight of solids
Sand drying bed	4	Dosing depth, cake solids	Area of drying bed, dry weight of solids
Landfill	–	Volume	Dry weight of solids, solids concentration

Establishing Feasible Processes. Preselection must clearly be distinguished from design as a separate step in the planning process. By preselection, the designer chooses main alternatives on which to concentrate more detailed evaluations. In practical planning, the preselection is often influenced by personal experience. This program, which is designed for domestic wastewater, or other biologically treatable wastewater with similar characteristics, should provide more quantitative information and, thus, improve the rational basis for the preselection of treatment processes. This may be of special importance for the choice between low and high technology options.

The program is designed to give the least cost solution. However, it is expected that for the preselection process, several scenarios will be simulated by varying key parameters related to process design and cost. The low running cost of microcomputers and the immediate availability of results makes microcomputers specifically suitable for these types of studies.

Process Design and Cost Estimates. Inflow concentrations, effluent standards, and the various process loadings influence considerably the size of treatment units and the resulting cost. Process designs are, therefore, performed separately for the various loadings of individual subprocesses, and depending on Biochemical Oxygen Demand (BOD) and Suspended Solids (SS) inflow concentrations and on effluent standards. The result of the process designs are characteristic dimensions, such as the volume of aeration basins or the required air flow for the activated sludge process. The process designs also give effluent concentrations, and quantities and concentrations of sludge. The latter are needed as input values for the sludge treatment processes. The process designs are performed according to common design procedures.

The characteristic dimensions, resulting from the process designs, provide the basis for cost estimates. The program contains a set of standard cost equations based on American sources. However, the cost for wastewater treatment varies widely, depending on local cost factors. Sensible use of the program, therefore, requires that the standard cost functions be checked for their validity for an individual project and modified if necessary. Experience shows that cost variations influence the absolute cost values, but change the relative costs to a lesser extent.

Selecting the Least-Cost Alternative. The program combines the described subprocesses into a large number of technically feasible treatment and disposal options. The complete treatment processes are developed and organized in the form of a decision tree and evaluated by means of a branch-and-bound algorithm. Figure 4-7 shows the main components of the decision tree, which contains six consecutive treatment steps, comprising between 3 and 54 alternative processses or process loadings. For example, step two

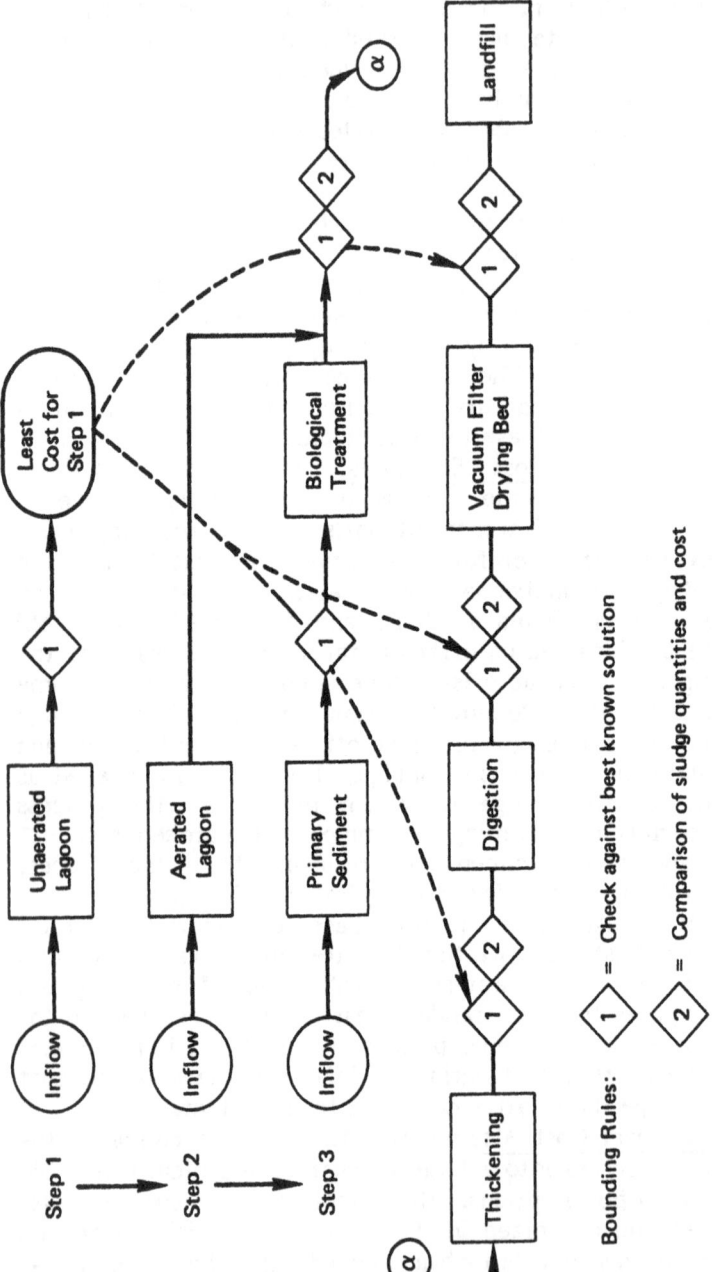

Figure 4-7　Logic of the branch and bound algorithm for optimization of wastewater treatment processes

comprises the biological treatment processes: activated sludge, trickling filter, rotating biological contactor, and various load or lagoon systems. In total, 265 process loadings are provided for the first three of these processes.

The branch-and-bound algorithm corresponds closely to the structure of the decision tree. The low technology options, which are unaerated lagoon systems, involve considerably fewer and simpler design calculations than do the high technology options. Accordingly, the low technology options are calculated first and compared against each other (Figure 4-7). The five alternatives with the lowest costs are stored and finally printed for further consideration. The least cost of all low technology options will be used as the elimination criterion in the following evaluation of high technology options.

When evaluating the high technology options, bounding rules are applied after each subprocess. Two different bounding rules are used. The first rule results from a comparison of low and high technology options. A high technology branch is terminated if the accumulated cost exceeds the least cost of the low technology options. The second bounding rule allows the termination of additional branches by comparing high technology options against each other. The sludge quantities produced at each stage are used as an additional criterion. If the cost of an alternative A is higher and the sludge quantities and concentrations not less than of an alternative B, alternative A will produce at the following sludge treatment stages no lower cost than alternative B. Alternative A, therefore, can be terminated. This second bounding rule is applied after biological treatment, thickening, digestion, and dewatering.

The decision tree of the optimization algorithm implicitly contains more than 39,000 alternative designs. They are checked for feasibility with respect to effluent standards and compared with respect to cost. The core storage requirement for the program is about 9K, the execution time in the range of minutes (IBM PC 5150, FORTRAN compiler). It is clear, then, that the capacity of microcomputers is sufficient for optimization programs for alternative treatment plant designs.

Setting Up the Design System. The more crucial points for practical applications are the standardization of design procedures by a computer program and the availability of reliable cost data. A great variety of design procedures is in use, but a design program can contain only a selection. The user must be able to modify the program if, in his judgment, it does not examine the most appropriate alternative.

The cost comparison of design alternatives demands reliable cost data for all the alternatives to be compared. Considerable work may be needed to prepare the database for an optimization program. It may be concluded that the capacity of microcomputers and the available optimization techniques are not the most restricting factors for practical application. The restricting factors are rather engineering aspects and the availability of cost data. For this reason, the user of the program is a key factor. The user must have the experience to ensure that all viable alternatives are considered and to provide realistic cost figures for comparison. In the hands of a suitable systems designer, the microcomputer becomes a valuable adjunct in expanding the number of alternatives that can be explored within a reasonable time.

Microcomputer Based Health Training, Bangladesh

A short-term microcomputer training program was carried out at the National Institute for Preventive and Social Medicine (NIPSOM) in Dhaka, Bangladesh (Frerichs, 1985). The Institute is a postgraduate educational institution responsible for public health training of physicians and other health professionals, and the program was part of a United Nations sponsored workshop on epidemiologic analyses using a microcomputer. All 15 participants received at least 2 hours of instruction for 10 days; five received an additional 2-3 hours per day of individualized instruction on the workings of the computer and spent 3 additional days working on an applications program.

The hardware consisted of an IBM portable PC with 640K of RAM and an 8087 math-coprocessor, an IBM expansion unit with a 10 MB fixed disc, and an Epson FX-80 printer. In addition, the appropriate transformer and voltage regulator were supplied, and a year's supplies such as ribbons, discs, and paper was provided. The IBM hardware was chosen after correspondence with counterparts in Bangladesh because repair service for the equipment was available in Dhaka. Because of local problems with dust and humidity, the equipment was installed in a closed, air conditioned room. A dedicated electrical circuit also was installed by local electricians.

Three package software programs were provided: PFS:File for entering data; BMDP PC Statistical Software, a comprehensive statistical analysis system; and PFS:Graph to illustrate the analyses in graphic form. All were chosen for ease of operation. PFS:File provided up to 32 data entry screens that could be formulated to replicate the pages of a questionnaire or interview form. The BMDP PC uses the same instructions and control statements as the mainframe version available in major universities and research centers throughout the United States and Europe. PFS:Graph was also clearly documented and easy to learn.

In addition to training the NIPSOM participants, the workshop had another purpose: to determine whether it was feasible to train technically sophisticated but computer illiterate professionals to use this technology in a relatively short time. All the participants understood the uses of the microcomputer for epidemiologic analyses and many were readily able to see applications both for themselves and their students. In addition, the more intensively trained group of five was able to enter information from one of their own survey data sets and to generate and print a series of tables using the BMDP program. The participants initially used two of the fourteen statistical programs provided. Once the power of the microcomputer was fully appreciated, the NIPSOM faculty requested additional statistical consultation to explain some more complex analysis techniques.

5

Applications in Energy

INTRODUCTION

This chapter focuses on the role of microcomputers in energy planning, analysis, design, control, and management. It does not catalog, but rather introduces a range of applications in which the microcomputer makes significant contributions to the nature and quality of analytical work in the energy sector in developing countries.

The range of potential applications of microcomputers in energy work is broad. Since the objective here is to present important applications for developing countries, the following selected uses will be discussed:

- National Energy Planning
- Planning and Design in the Electricity Sector
- Engineering Design and Analysis of Energy Systems
- Process Control and Energy Management
- Energy Resource Development.

These uses are representative of microcomputer applications that are crucial to planning and development efforts. Further, the range and type of microcomputer applications in these areas provide insights as to their use in nonenergy but related areas as well.

APPLICATIONS REVIEW

National Energy Planning

The application of microcomputers in energy sector planning and analysis is no accident. Their use reflects a decade of energy planning activities in the developing countries as well as in engineering, design, and project management primarily in the industrialized countries. For approximately five years, microcomputer-based systems have both simplified the tasks of data analysis and modelling and brought computing power to project work in developing countries.

During this same period, significant strides were being made in the use of microcomputers in business, industry, design offices, and universities in Europe, the United States, and parts of Asia. This led to widespread adoption of microcomputers for process control in manufacturing; computer-aided design in engineering, manufacturing, and architecture; and for use as a tool for analysis by university students. The spillover for use in developing countries was a natural consequence.

The history of their adoption begins with energy planning during the late 1970s. The U.S. Department of Energy (DOE) sponsored developing country energy assessments in Egypt and Peru in 1976-1977. Similar studies followed in Argentina, Portugal, and Korea. In addition to these general assessments, a number of sector-specific studies focused on renewable resources in the Sudan and other African nations, industrial energy conservation in Tunisia, energy investment portfolio selection in Morocco, energy information systems in the Dominican Republic, and fuelwood plantations in the Philippines (Munson and Palmedo, 1983). In addition, the Swedish government, through the Beijer Institute, conducted energy demand studies in Kenya; the French did the same in Asia. The Canadian International Development Agency (CIDA) was also active in sponsoring energy sector projects. The European Economic Community (EEC) also supported energy assistance programs, and a UNDP-World Bank energy assessment program was also initiated. These efforts reflected the critical balance-of-payments problems associated with high prices for oil in oil-importing developing countries. This problem still exists, and the longer-term need for sound investment plans for the energy sector continues.

Quite by coincidence, the Apple computer was born in the late 1970s. At the same time, oil price rises were being felt, and developing countries naturally turned to energy sector studies. But it was not until the period 1978-1980 that microcomputers were first applied in energy sector work in developing countries. A few microcomputers began to appear in several countries; for example, a Hewlett-Packard microcomputer was introduced in Nepal's hydrology studies.

The microcomputer comes into use in three ways in the energy sector of developing countries:

- Demand Push--where within a given energy sector project there is need to cope with "number-crunching" that begins to overload the staff. Generally, a machine is first borrowed to carry out the activity, leading to demand for owned or dedicated microcomputer equipment.
- Top Down--where the conception of a project, from its inception, is sufficiently complex that it must rely on microcomputers. Upper-level management is directly involved in structuring a project that inherently involves commitment to microcomputer support.
- Educational Complement--within energy training seminars, where there is the desire to go beyond classroom experience and into useful applications. In professional level education, the concepts may be new but usually build on a solid foundation of knowledge gained beforehand. The microcomputer facilitates extensive work with actual data and the utilization of analysis skills at the professional level (such as determining energy balances, net present values, and benefit-costs, all of which are quite difficult to do in detail without microcomputer equipment).

The introduction of microcomputer enhances energy planning activities in three distinct ways. First, the microcomputer allows stretching the limits of analytic work in the early stages of energy sector activity. The microcomputer is perceived as powerful and capable of analytic work that might formerly have been considered too difficult. For example, when the Morocco energy investment study commenced, the idea of including uncertainty in the analysis was interesting, but its implementation was computationally difficult. The existence of microcomputer support allowed attempting the more challenging problem, coming closer to the ideal in decision analysis, and probing a larger number of alternatives.

Second, the microcomputer lends a perception of quantity and quality to energy sector analysis. On one hand, there is a legitimate belief that the microcomputer forces users to become more organized in their work and develop more extensive sources of data. At the same time, there is an assumption, though perhaps it should only be hope, that the microcomputer encourages the analysis of more possibilities or cases with respect to any given decision. As a result, with the microcomputer, the availability of more information leads to better decisions.

Third, in contrast to the mainframe alternative, analytical work performed on a microcomputer brings the computational process closer to the analyst. In addition, through the use of graphics, the computational and analytical processes are also made clearer to the decision maker.

To develop an understanding of the role of the microcomputer in energy applications and the problems that attend its introduction, three case studies are examined below. In reviewing their history, conclusions can be drawn regarding the benefits and liabilities of using microcomputer equipment in these settings. The character of interaction with microcomputers and the experience of the participants in these examples is quite similar to that of the individuals involved in the other applications such as health or agriculture.

Table 5-1 lists a selection of energy sector planning and development models used recently (Munson et al., 1983). For completeness, a number of models that were prepared for U.S. energy sector analyses are also included, since these models might become available in modified form for developing country applications. Indeed, in a number of areas, such as economic projection, demand forecasting, and energy supply-demand balances, this translation of the U.S. experience into tools for developing countries has already occurred. In other cases, specific tools have been developed with the full participation of developing country economic and energy planners.

Planning and Design in the Electricity Sector

Context and Software

Microcomputers have been used for planning and design in the electric sector in several developing countries. However, software for this purpose, although widely used, is not widely available. For example, microcomputers are used for energy-water resources planning in Nepal, for both technical and policy analysis, project design, and financial evaluation. The power sector plan for Bangladesh involved substantial analysis on an IBM-PC. Electric sector planning in Tanzania will rely on microcomputer analyses. The software for these applications includes load flow, transient analysis, water balance, and other technical and economic programs. However, these programs required extensive experience and training for the engineers and economists who will use them. The magnitude of money and personnel investment to utilize such software effectively will probably limit their use for the immediate future.

The development of utility planning models itself is an expensive, detailed undertaking that requires sector-specific experience and knowledge. Historically, only sizable utilities in developed countries, architectural and engineering firms, and large multiprogram consulting companies have had the experience necessary to develop this software. Those companies often consider such

Table 5-1 Energy Planning and Management Models

Model Type	Mainframe Models	Microcomputer Models	
		Description	Where used
Energy Databases	1) Energy, Modelling Database (EMDB), Brookhaven National Laboratory 2) Energy Supply Planning Model, Bechtel Corp. 3) Energy Balance Database (EBDB), Argonne National Laboratory.	1) ALEIDIS System SUNY	1) Dominican Republic
Energy Supply and Demand Balances	1) Argonne Energy Model, ANL 2) Reference Energy System, BNL 3) LEAP, FSRG/Beijer (Kenya)	1) Argonne Energy Model, ANL 2) Energy Systems Planning Model, DSI, under development 3) RESGEN; IDEA, INC. 4) Enerstat, under development, proprietary 5) LEAP 6) RMA model	1) Tunisia, Sudan 2) Morocco 3) Uruguay, Haiti, Thailand, Sri Lanka, Indonesia, Dominican Republic 4) Sudan 5) Kenya, Angola 6) Tunisia, Indonesia
Economic and Pricing Studies	Various	1) E/DI Energy Pricing Analysis Model (EEPAM), E/DI 2) Energy Finance Assessment Model (EFAM), IDEA	1) Somalia and Liberia 2) Tunisia, Sri Lanka, Dominican Republic

Table 5-1 Energy Planning and Management Models (continued)

Model Type	Mainframe Models	Description	Microcomputer Models — Where used
National Energy/Economic Analysis	1) Midterm Energy Forecasting Systems (MEFS), USDOE) 2) A Macroeconomic Model of Inflation and Growth in Korea, Korea Development Institute 3) BEEAM, BNL	1) Energy-Macroecononomic Accounting Framework, Sri Lanka Ministry of Power and Energy	1) Sri Lanka
Sector Specific	Several examples: National Coal Model; On-shore Oil and Gas Supply Model; Aggregate Refinery Model; Solar Heating and Cooling Model; Residential Energy Use Model; OR Industrial Model (ORN); Electric Power System Development Model; Technology Assessment Model (SRI); Oil Market Simulation Model	No comparable public micro models 1) Fuelwood/Forestry Model, The Futures Group 2) Various proprietary oil and gas systems models 3) BMASS Fuelwood Model, IDEA, INC. 4) ISPLAN Electric Sector Model, IDEA, INC.	None 1) Several African countries 2) No data 3) Sri Lanka 4) India, Thailand
Energy Conservation Analysis and Models	See Table 5-5	1) Various models and databases	1) Tunisia, Togo

Table 5-1 Energy Planning and Management Models (continued)

Model Type	Mainframe Models	Description	Microcomputer Models	
				Where used
Optimization Models	1) Brookhaven Energy System Optimization Model, BNI; 2) Regional Industrial Supply Model, DRI; 3) Project Independence Evaluation System, DOE 4) Systems Europe Optimization Model	1) REFINE Refinery Model IDEA, INC.	1) Sri Lanka	
National Investment and Financing		1) Energy System Planning Model, DSI; 2) Finanace and Cash Flow Analysis Model, IDEA, Inc. 3) Energy Investment Assessment Model, E/DI	1) Morocco 2) Haiti, Dominican Republic 3) Dominican Republic	
Input/Output Models and Macro-economic Models	1) Energy Disaggregated Input/Output; 2) Generalized Equilibrium Model 3) Wharton Annual Industry Forecasting Model	1) No comparable models 2) Colombia 2000	1) None 2) Colombia, Uruguay, Bangladesh	

software proprietary, using it as a cost-cutting and marketing tool for consulting and development work.

To a certain extent, these conditions are changing, and in Table 5-2 examples of computer programs are listed, but not all of them are available on the microcomputer (Munson et al., 1983). Early on, the U.S. Department of Energy sponsored the development of many models, including nonproprietary electric utility models for general and specific purposes, which are now emerging as market-makers. An example of how microcomputers can be used in electricity sector distribution system design and planning is discussed in the following section.

Graphics for Electricity Sector Management

Traditionally, electric utilities have recorded changes to their systems on hand-drawn maps stored in bulky files. Utilities in both developing and developed countries are seeking more efficient ways of updating and storing maps and records. Recent advances in microcomputer hardware and software in computer graphics makes this possible.

Computer-aided graphics is a term used to describe computer-generated drawings. Drawings are digitized, that is, converted from visual form, and stored on a computer. Since existing maps are generally either old or of poor quality, new maps usually need to be drawn if analytical work is to be undertaken. Aerial photographs can be used as a starting point. Individual points on a map can be located vertically to establish elevations by obtaining photos from different vantage points and combining the images under a stereo plotter, producing a three-dimensional effect.

The photos are then digitized and entered directly into a computer by placing them on a digitizing board containing closely spaced wires and typically picking off points with an electronic pen or pointer. This is much simpler than keying in X-Y coordinates of various features and photos. A base map reference system can be created this way. The scale will depend on the specific application and the area to be covered. Specific information can be added once the reference map has been created. This might include pole heights, voltages, conductor sizes, transformer information, load, and generation characteristics.

Specifying such a system is still more art than science, since both equipment and software are a function of the exact application envisioned. Few complete turnkey systems are available, and a trial period is therefore recommended. Points to be considered before purchase are the map sizes and scale desired and the quality and type of database that will be used to generate the maps. Map and database use must be defined, and symbols agreed on and developed. The system must be capable of being used on a daily basis and be expandable to meet future needs.

Table 5-2   Examples of Electric Planning Sector Models

Model	Source/Vendor
Wien Automatic Planning System Model (WASP Versions, I, II and III). Determines optimal national electric generation expansion program using system reliability and cost.	International Atomic Energy Agency (IAEA) and Argonne National Laboratory (ANL-minicomputer)
Reliability Computation Model (Relcomp.). Evaluates the reliability and economic performance of alternative generation expansion plans (nonoptimizing).	ANL (microcomputer with extended memory)
Energy Supply Planning Model (ESPM). Determines capital and operating cash flow requirements for given supply system configuration. Also estimates labor and material requirements for construction and operation of facilities.	ANL (Apple II)
Load Flow, Transient Analysis, Water Balance Modules.	ACRES (IBM/PC)
WASP (for IBM-PC). The Wein Automatic Planning System has been adapted to run on the IBM-PC. In addition to WASP features above, adaptations include marginal cost pricing and economic despatch, renewable technology interaction with the grid, and other features.	ENTEK, Inc. Setauket, NY (IBM-XT)

Computer-aided graphics, being adopted by utilities in the United States and elsewhere, allow analysts and planners to study proposed and existing grid components through computer-enhanced graphics. By combining both physical site characteristics with non-graphic data, such as income or family size, they can enhance the planning process. Selecting such a system is a complex task and must be carried out with a great deal of study, but the benefits will soon be evident.

Engineering Design and Analysis of Energy Systems

For energy systems to operate economically, they must be designed to provide the greatest energy or work output per unit energy input. The efficiency of today's thermal processes (such as industrial and power boilers, furnaces, heaters, transportation systems, and metallurgical and petrochemical processes) is much improved due to better materials, design, and energy recovery. Brought about primarily by the high cost of fossil fuels, improvement proceeded incrementally as old equipment was replaced or retrofitted. To remain competitive, manufacturers of thermal equipment have provided more energy efficient designs.

The engineering of such equipment often requires bench scale testing and the development of mathematical models used to optimally design commercial scale equipment. Before the introduction of microcomputers, data and models were analyzed on mainframe computers, which were available only to manufacturers in industralized countries. The microcomputer has made it possible for small enterprises in developing countries to manipulate and analyze demonstration plant test data and alternative designs.

Design software for a variety of engineered systems is currently entering the commercial marketplace. Many of these packages are being offered by small firms or private consultants. They save countless hours of programming time but should be used with caution. Often it is not clear what design relationships have been used or whether if they were programmed properly. When planning to use such packages, it is best to consult with others who have used them or to check the computations by traditional methods.

Automated Design

The design of engineered systems by microcomputer has become standard practice; it is popularly known as computer-aided design (CAD), computer-aided engineering (CAE) or computer-aided design and drafting (CADD). Examples include three-dimensional piping layouts for the process industries, digitized geographical analysis for topographical use, and two-dimensional and three-dimensional structural analysis of buildings. Perhaps the most interesting feature of CAD is the way it permits the engineer to interact with his design through graphic display.

Energy Efficient Thermal Equipment Design

The analysis of system components such as pressure vessels, turbines, pumps, electrical controls, and building energy systems are all necessary for determining the optimal design of a complete energy system. Programs are commercially available for determining the optimal design based on energy constraints. These programs are distinct from the graphic design programs discussed previously in that they are based on engineering performance—that is, temperature, speed, pressure, and load bearing ability. Usually a series of equations is developed and solved simultaneously, yielding the optimal design parameters based on constraints defined by the designer.

These programs generally provide solutions based on mass and heat balances or the most economical design. These balances will help the engineer determine critical dimensions and control parameters for operation. Many of these programs were developed on mainframe computers, but most of them have been downloaded (or adapted) for microcomputer operation.

The thermal design of components, such as pipe insulation, boilers, furnaces, and burners, can be carried out with a microcomputer using commercially available software. Digital Equipment Corporation has a Software Referral Catalogue, which lists a variety of chemical and mechanical engineering software useful for such purposes. The Engineering Software Exchange, published by CAE Consultants Inc. (Yonkers, NY), is another source of information.

Unfortunately, much of this software is proprietary and remains in the hands of its developers, usually private firms whose role is to provide design services. However, in developing countries these programs can be generated by university groups or local design firms to fit local circumstances. There are many low cost design aids available, such as specialized books containing programs for solving mathematical equations, and commercial software such as spreadsheets, which were orginally developed for business but which can be used for engineering analysis.

Many of these programs are also in the public domain and are available from a variety of U.S. government sources, research laboratories, and universities. Table 5-3 lists some of the available software.

Table 5-3   Examples of Design Software for Thermal Equipment

Program	Vendor
Boiler Efficiency	Consulting Engineering Services
Heat Exchanger Design	Hemisphere Publishing
Combustion and Steam	Software Systems
Power Plant Simulation	Physical Sciences Inc.
Design of Process Components for Petrochemical Process Industries	Dynacomp Inc. Engineered Software Kelix Software L.N. Engineering Magellan Systems PSI Systems Technobyte
2D Design and Drafting	Auto Desk Inc. MCS Inc. United Networking Systems T&W Systems Inc.
3D Design and Drafting	Computer Aided Design Kern International

Solar System Design

Recent emphasis on solar energy as a substitute for fossil fuels has encouraged the use of microcomputers for design. In the United States, there are software design aids for solar water heaters, greenhouses, passive solar homes, and photovoltaic systems. Some examples are given in Table 5-4. Most of this software can be applied to designs in developing countries.

Computer modelling of solar systems allows the analyst to simulate a variety of meteorological conditions and determine the thermal response of a particular design configuration. For example, TSWING is a solar design package available from Solarsoft that allows an Apple computer user to design a living space, such as a greenhouse, and determine its thermal characteristics throughout the year (See Figure 5-1).

Table 5-4   Examples of Solar Software

Program	Vendor
TROMBE: analyzes thermal storage devices	The Bickle Group
SOLDHW: calculates energy savings from solar-hot water	The Bickle Group
SOLAR COLLECTOR: F-Chart and Economic Analysis	McClintock Corp.
PASODE 1: calculates auxiliary heating requirements	Lende-Parker-Michaels
IMPSLR: calculates solar savings fraction and auxiliary heating requirements	Princeton Energy Group
SOLAR STAIRCASE: calculates transmission, absorption and reflection	Norman B. Saunders
SUNPAS AND SUNOP: calculates solar gain, heat losses, and requirements, etc.	Solarsoft
TSWING: calculates temperature profiles for various design configurations	Solarsoft

118

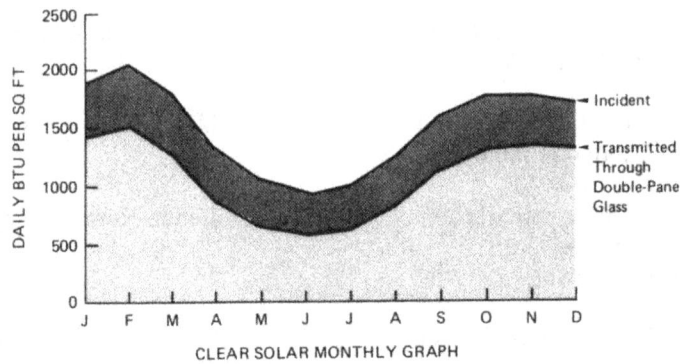

LATITUDE: 42   AZIMUTH: 0   TILT: 90
MONTH 1

HOUR	SUN ALTITUDE	SUN AZIMUTH	SUN ANGLE OF INCIDENCE	DIRECT RADIATION	DIFFUSE RADIATION	REFLECTED GROUND	TOTAL SOLAR GAINS	SOLAR GAINS IF A SHADING DEVICE USED	SHADING FACTOR (% OF WINDOW SHADED)			OPTIONAL SIZE WINDOW (1 SQ. FT.)
							INCIDENT BTU				TRANSMITTED BTU	
HR	ALT	AZM	INCID	DIR	DIF	REF	TOTAL	SHADED	S.F.	TOTAL	SHADED	WINDOW
8	7	55	55	68	3	2	74	74	0	54	54	54
9	15	44	46	159	7	7	173	173	0	130	130	130
10	22	30	37	213	8	12	233	233	0	176	176	176
11	26	16	31	244	8	14	267	267	0	204	204	204
12	28	0	28	254	8	15	278	278	0	213	213	213
13	26	16	31	244	8	14	267	267	0	204	204	204
14	22	30	37	213	8	12	233	233	0	176	176	176
15	15	44	46	159	7	7	173	173	0	130	130	130
16	7	55	55	68	3	2	74	74	0	54	54	54
							1772	1772		1339	1339	1339

Figure 5-1   Typical output for a solar design living space. (Milstein, J. "Use Your Computer As A Solar Design Tool," Popular Science, December, 1982)

The program begins by asking for the x-y-z coordinates of the planned structure and generates a three-dimensional representation on the screen. The program then computes hour-by-hour temperatures, allowing an analysis of different thermal masses and glazing. A subroutine called SOLGAIN will compute incident and transmitted radiation for any latitude and shading configuration. A variety of such tools is available from commercial vendors, as well as through U.S. government sources.

The microcomputer and new software make solar design feasible on a local basis in developing countries. In fact, several countries have solar energy research institutes and university groups that can easily develop and adapt such models to fit local conditions. Empirical data accumulated at these centers can be used as a database for modelling and design. The data needed usually include solar insulation, cloud cover, wind direction, speed, and temperatures. The development of solar software should be encouraged in those areas where it has application, since both construction materials and environment are site specific.

Industrial Energy Conservation

Until recently, industrial energy conservation programs in developing countries were the focus more of talk than action. This was, perhaps, reasonable, given that relatively little was known about the efficiency of industrial processes in these countries. Following initial national energy assessments and planning efforts, however, it became apparent that potential savings from energy conservation would be significant. Therefore, numerous assistance efforts are now targeted to that end.

Many energy conservation software packages are available commercially or at cost through nonprofit organizations. These packages satisfy many of the primary needs of industrial energy conservation managers. Many of these packages result from industrial energy conservation efforts in the United States and Europe, where significant reductions in per unit energy consumption have occurred in industry. Table 5-5 shows a number of programs useful for specific and general energy conservation applications.

Table 5-5   Examples of Software Used in Energy Conservation
Analysis and Planning

Program	Vendor
STEAMPROP, thermodynamic properties of steam; HEATLESS, computes the heat loss from pipes and flat surfaces; CURVFIT; DFCECA, discounted cashflow analysis for energy conservation alternatives; PIPELOSS, calculates head loss and power consumption.	Center for Energy Studies University of Texas at Austin.
HVAC Energy Consumption and System Simulation; Lighting Design; Residential Cooling and Heating; Heating Fuel Cost; Commercial and Residential A/C Group; Commercial Cooling and Heating Load--ASHRAE.	McClintock Corporation MC Programs
ENERCOM, provides utility companies with a system to evaluate residential energy efficiency.	Enercom, Inc.
PROCESS, general purpose flowsheet for the calculation of mass and energy balances.	Simulation Sciences, Inc.
TRACE, assists building and systems designers in comparing operating efficiencies and costs of alternative A/C systems.	Trane Air Conditioning Economics
EFACT, analysis of the potential savings obtained by implementing alternative control strategies.	Johnson Controls, Inc.

Table 5-5   Examples of Software Used in Energy Conservation
Analysis and Planning (continued)

Program	Vendor
ECD, energy conservation design; WONDER-2, estimates yearly energy consumption; EMCS, determines energy and cost savings from alternative HVAC control strategies; EFFUFAC, models head gain/loss through a roof or wall.	Bickle Group computer programs
Boiler Plant Performance Analysis; Boiler Turbine Plant Efficiency	International Micro Software Catalogue, Microbits Ltd.
Commercial Load Estimating Operating Cost Analysis Lifecycle Cost Analysis	Carrier, Inc.
Plant Level Economic Analysis of Energy Conservation	Hagler-Bailly

Munson et al., "The Use of Microcomputers in Energy Planning for Developing Countries," Final Report, Office of Energy USAID, 1983.

Industrial energy conservation is a focus of energy planning in Sri Lanka. Hagler-Bailly, for example, has developed an analysis of a boiler retrofit in a textile plant with software for the IBM-PC. The analysis is neither for a single boiler option nor a comprehensive look at the entire energy sector. Instead, it proposes an energy conservation program consisting of a number of options within a given textile plant and then determines economic feasibility. Most important, the procedure reports monthly cash flow and other financial data required to obtain commercial bank loans for the project. An industrial energy audit program to create both an initial database and develop local skills to analyze industrial energy consumption patterns is underway at the Industrial Research and Development Institute of Central America (ICAITI). The program includes in-plant energy audits and technology transfer workshops. Energy conservation analysis software developed over a number of years at the Georgia Institute of Technology is being considered for application via microcomputers.

At the plant level, many engineering calculations are performed routinely by energy managers and engineers The close user interaction provided by personal microcomputers makes programming for these calculations convenient and accurate. Common examples are:

- Furnace efficiency based on stack gas analyses and temperatures, taking into account fuel characteristics
- Boiler efficiency, same as above but also including the effect of blowdown and condensate return
- Heat losses from unit operations, such as boilers, process vessels, cement and brick kilns, and pipes, and the effect of adding insulation
- Fuel combustion calculations, to calculate stoichiometric air requirements for different fuels, stack gas volumes calculated from fuel compositions
- Heat exchanger calculations and sizing of heat recovery systems
- Heating, ventilating, and cooling loads for offices and commercial buildings.

Some software packages are available commercially for these types of engineering calculations. The programs listed in Table 5-5 are examples.

Some of the most useful industrial energy analysis work relies on VisiCalc and LOTUS, the most popular of the "electronic spreadsheets." These can be adapted to analyze a variety of problems that can be expressed in matrix or tabular form. For example, energy users often have to project energy demands over a period of time for which annual growth rates are specified.

Suppose sector specific annual growth rates are believed to be subject to considerable variation: what would industrial demand become if the construction materials subsector were actually to grow at 10.5 percent per year? By setting up the "matrix" calculation, using the spreadsheet program, the user can simply amend the subsector growth rate to 10.5 percent, and the computer will automatically recalculate the projected demands and other dependent figures within a few seconds. Other key data can be changed by the user, and the sensitivity of the projections to these changes can be observed immediately.

Another major type of data-handling activity is the storage and retrieval of large masses of data. An example of this application to energy conservation is the collection and analysis of plant-level energy data carried out at 60 factories in Tunisia by E/DI Europe. Visits were made to all the plants and a questionnaire completed for each by the survey engineers. Data collected included energy consumption figures for each energy form and corresponding production figures. The analyses carried out using this information and Versaform, a product of Applied Software Technology, included a review of consumption for different subsectors and comparisons of energy efficiency with international figures for corresponding manufacturing activities. Using the same Versaform software, it was possible to prepare a number of useful reports by searching the data files in different ways.

Process Control and Energy Management

The use of mini- or microcomputers for process control has become standard practice in industry in developed countries. Microprocessor-based control devices are used for a variety of work station operations (such as machining and assembly in manufacturing), for process regulation in petrochemical plants, and for operating biomedical instruments. Clearly, many of these applications are in energy-related fields. Since microprocessor-based controls are relatively inexpensive compared with the equipment they control or with previous techniques using relays, they are suitable for application in process control in a variety of energy saving roles in developing countries.

Programmable controllers (PCs not to be confused with personal computers) or programmable logic controllers (PLCs) are used in a variety of industrial processes. In power plants, for example, they are used for coal or oil handling, pollution control, water treatment, boiler control, combustion and air flow, feedwater heating, and to control ash handling. They may be distributed, that is, modulating a discrete set of variables associated with one step or unit operation in a process, or they may be centralized, that is, controlling an entire process. The trend is toward distributed systems for both technical and economic reasons.

PLCs today perform functions that were traditionally left to relays and solid state logic systems. These devices, by comparison, were slow and cumbersome and subject to frequent failure. In most developing countries, process control may be manual or, at best, may utilize outdated technology such as mechanical systems—governors for speed control, for example. The new microprocessor-based technologies can replace many of these outdated control components. Usually, for the cost of one year of maintenance of such mechanical control elements, a microprocessor-based controller can be substituted.

Two examples of process control and energy management are given in a subsequent section to illustrate the techniques available: (1) a data acquisition system, for which a microcomputer with both custom and off-the-shelf software is used; and (2) an in-depth example of building energy utilization where the micro-computer is used as an engineering tool to design more energy efficient buildings.

Energy Resource Development

Evolving energy strategies in developing countries focus on a variety of problems and opportunities. One strategy emphasizes the development of indigenous and renewable energy sources as alternatives to imported fuels and as more appropriate to the dispersed, nonintensive nature of agricultural and domestic energy demands.

The large range of such alternatives includes active and passive solar systems, micro, mini, and small hydropower projects, agricultural and animal wastes, and of course wood. Of these, fuelwood use has received fair attention because its use accounts for a significant fraction of current consumption and, consequently, has contributed to the process of deforestation and desertification.

Evaluating the development of these resources involves such considerations as the distribution and energy intensity of the resource; selection of the most appropriate recovery and conversion technology; comparison with the specific need it will fill and/or technology it will offset; evaluation of the costs and benefits; and, ultimately, an analysis of the effect of using these resources on the national energy and economic systems. Table 5-6 presents a sampling of renewable energy software (refer to Table 5-4 for solar software).

The availability of these models is directly related to the historic demand for such planning tools in developed countries. It is natural, therefore, that solar and related models dominate because there is a ready market for their application. It is also to be expected that more forestry and fuelwood models will become available. Forestry problems have been the focus of significant work over the past five years, and the use of fuelwood is one of the major factors contributing to the decline of forests in many countries.

Wind, biomass, solar, photovoltaic, cogeneration, and district heating microcomputer-based models spanning the variety of needs and opportunities are rare. The availability of broader, decision-oriented models would make a major contribution to energy planning in developing countries. These alternatives are among the most important opportunities for developing indigenous energy supplies and alleviating continued dependence on imported fossil fuel supplies.

Table 5-6 Renewable Energy Systems Models

Energy Source	Models	Source
Forestry/Fuelwood	LEAP	Energy Systems Research Group
	FUELWOOD	The Futures Group, Inc.
	EEFAM (E/DI Energy Fuelwood Analysis Model)	Energy/Development International
	BMASS	Sri Lanka Fuelwood Model
	Total Resource Base (TRB)	Dartmouth Resource Policy Center
	Fuelwood Supply-Demand	Energy Environment Engineering
	BIOCUT	Oak Ridge National Laboratory
Wind	SYSWIND	CIED, ANL
Petroleum	Integrated Petroleum Planning System	Scientific Software-Intercommunication
Small Hydropower	Technical Design and Evaluation	InterAmerican Development Bank
	Prefeasibility Evaluation of Cost and Capacity, Average Energy and Economic Return	University of Minnesota/ St. Anthony Falls Hydraulic Laboratory
	Hydrologic Evaluation and Prefeasibility Investigations	U.S. Department of Energy/ E.G. & G. Idaho

Petroleum Exploration and Development

The problems of seismic interpretation, data evaluation to estimate the performance of wells, and other aspects of petroleum exploration and field development generally require extensive computation and analysis. Such mathematical computations have been beyond the scope of microcomputers until recently. The newer 32-bit processors with graphics capabilities are now used for such purposes. Scientific Software, in conjunction with Hewlett-Packard, offers an integrated package of petroleum exploration and production programs running on the HP9000 workstation. The software includes mapping, well testing, log analysis, and production history, all in an interactive programming environment. Although the equipment and programs are expensive, the package can represent an impressive array of skills and knowledge in the petroleum field in the hands of developing country geologists and engineers.

Fuelwood Models

Increasing demand for fuelwood and wood energy feedstocks has placed severe pressure on forest resources in both developed and developing countries. Energy plantations are receiving considerable attention for significantly increasing the size and productivity of the wood resource base. Fast-growing trees, short-rotations (3-10 years), dense tree spacings (1-6 square meters per tree), extensive weed control, and in some instances fertilization to increase productivity and/or to replenish soil nutrients characterize this energy concept. These are generally broadleaf trees that coppice or resprout after cutting to regenerate succeeding biomass crops at little or no additional cost.

Potential and existing applications of this concept include intensively managed plantation systems in moist temperate regions, systems with low levels of management in arid and semiarid zones, and high productivity systems in tropical and semitropical areas. In each case the social and economic appraisal issues are similar. First, how do the economics change from one wood energy plantation site to another? Second, which species and management alternatives offer the greatest net return per hectare? And third, in which areas of plantation design do current uncertainties about biological and economic relationships lead to the greatest financial risk?

To help analyze the variety of alternatives for production, management, and use, several microcomputer-based models have been developed. These models are the result of fuelwood surveys and analyses that were performed during the late 1970s in several developing countries (National Academy of Sciences, 1980). These surveys also provide the basis for fuelwood production projects currently underway. Two models in use are BIOCUT, developed at the Oak Ridge National Laboratory, and the Futures Group's Fuelwood Model. BIOCUT is primarily an investment economic model and is described in a later section.

The Futures Group's Fuelwood Model forecasts fuelwood demand, supply, and deficits based on population projections. This work, funded by AID during 1982, was developed on Apple microcomputers and can be expanded to include other renewable energy sources, such as agricultural residues (Cole and Edelman, 1983).

The model determines demand on the basis of per capita energy consumption of wood and charcoal, by both urban and rural populations. These values can be modified to account for efficiency in the conversion process. The supply is computed on the basis of yield and area of various categories of woodlands. These figures determine the rate at which wood can be removed without continued depletion.

The model is designed as a framework for fuelwood policy analysis and planning. It therefore can assist with funding allocations and fuel substitution. Since it is primarily a demand/supply model, it cannot compensate for population density or variable biomass yields due to weather conditions. It is also not detailed enough to provide in-depth economic analysis for feasibility studies.

Small Hydropower Evaluations

Small hydropower is both a very old and a very new energy resource. As one of the earliest sources of mechanical and electrical energy, hydropower was a major foundation of industrial development through the ninteenth and early twentieth centuries, only to fall into neglect in the era of inexpensive oil and nuclear power. These early hydropower facilities are now considered small in comparison with the power output of a project on the scale of Tarbella, Hoover, or Itapu.

The escalation of fossil fuel production costs over recent years, however, has prompted a return of small hydropower development throughout the world. In North America and Europe, small hydropower is a source of power that can come on-line rapidly (within 2 years rather than within 6 to 10 years), thereby reducing the risks associated with long-term projects, and can often compete in cost effectiveness with thermal power plants. Many hydropower sites that were abandoned in the 1960s in favor of thermal plants are currently being rehabilitated or retrofitted with new equipment.

The case for small hydropower is even stronger for developing countries. Small hydropower can utilize the indigenous resources in management, engineering, and construction to a greater extent than can large projects. Equipment manufacturing can eventually follow. In addition, small hydropower can bring electricity to remote regions where no national electric grid exists. Very small projects, microhydro, can utilize the construction skills of local village residents.

Small hydropower may be regarded as a facility with a total capacity of less than 30 MW. Minihydropower has between 100 kW and 1,000 kW (1MW); microhydropower usually has a capacity of less than 100 kW. In this book, minihydropower and microhydropower are included in small hydropower.

Small hydropower schemes can either be single purpose, such as the production of electricity, or multipurpose, where energy production is one aspect of the total use of the facility. Multipurpose facilities include those in which hydropower is developed in conjunction with irrigation, flood control, navigation, and water supply.

Prefeasibility studies for small hydropower typically require a great deal of repetitious calculation to determine the power and energy available at a site. In addition, there is usually insufficient time available to obtain detailed estimates on equipment and construction costs. Instead, generic cost curves are normally used. These approximations are sufficiently accurate for the prefeasibility investigations. They also make the prefeasibility process especially adaptable to standardization with a computer program.

Several studies and microcomputer programs have been developed to aid the analyst in determining the technical and economic feasibility of small hydropower projects. For example, the World Bank has developed a methodology for determining the hydropower potential of a specific region based upon sampling and statistical inference (World Bank, 1984). This methodology is being computerized by the Tudor Engineering Company of San Francisco.

St. Anthony Falls Hydraulic Laboratory in Minneapolis, Minnesota, has developed a set of microcomputer programs written in BASIC that can be used for prefeasibility investigations (Gulliver, 1984). Called HYFEAS, it uses flow and head data, information on existing and proposed hydroplants, operational constraints, standardized cost curves to estimate plant capacity, annual energy production during peak and off-peak hours, and equipment and plant cost, at increments of 5 percent on the flow duration curve. HYFEAS will also perform an economic analysis for public financing if desired. HYFEAS is a user-friendly, menu-driven program developed on an IBM-PC microcomputer.

The flow duration curve gives the percentage of time a given flow has been equalled or exceeded for the period of record. A daily flow duration curve is developed by taking all daily flow data and ranking them according to discharge, regardless of the sequence in which they occurred. The percent of the daily flow measurements equal to or greater than a given flow measurement is then calculated. This flow measurement is then plotted against the corresponding percent exceedance (see Figure 5-2 for a computerized version).

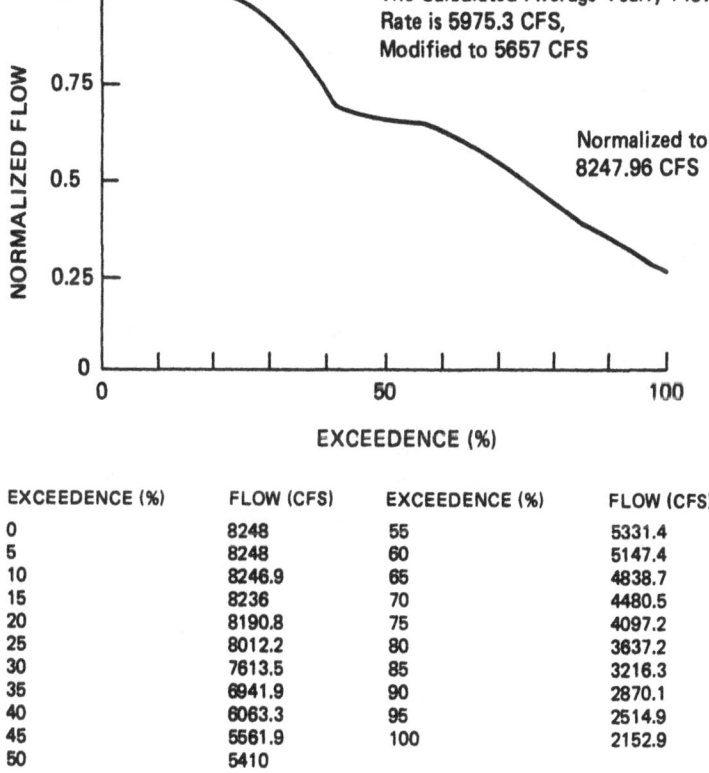

EXCEEDENCE (%)	FLOW (CFS)	EXCEEDENCE (%)	FLOW (CFS)
0	8248	55	5331.4
5	8248	60	5147.4
10	8246.9	65	4838.7
15	8236	70	4480.5
20	8190.8	75	4097.2
25	8012.2	80	3637.2
30	7613.5	85	3216.3
35	6941.9	90	2870.1
40	6063.3	95	2514.9
45	5561.9	100	2152.9
50	5410		

Figure 5-2   Computer output for a flow duration curve

The average power and annual energy produced by the hydro-plant is found by constructing a "power duration curve" from the flow duration curve and corresponding net head and plant efficiency information. The area under the power duration curve will give the average annual energy production. This procedure can be used to choose sites according to criteria such as the lowest installed cost or a positive cost-benefit ratio.

The U.S. Department of Energy has also funded a computerized reconnaissance analysis (HCR) made up of three programs, HYDRO-CALC, PAPER-ECON, and HYDRO-ECON (Broadus, 1981). HYDRO-CALC performs preliminary engineering calculations for a given hydroelectric site. Either the river flow exceedance curve or monthly flow is used as input. After the flow characteristics have been determined, various configurations of turbines can be evaluated and the amount of energy produced and the capacity factor can be computed. Characteristics for six types of turbines are provided in the program, and up to three turbines may be specified at the site.

PAPER-ECON is a program designed to provide rapid economic analysis of a site with nominal input to determine whether further reconnaissance or feasibility work should be performed. The minimum input normally required is the average flow rate at the site, the available head, the current date, and the site location. The program uses built-in equipment costs, benefit estimates, and site economics. However, most of the default assumptions may be overridden if specific information is available. HYDRO-ECON is similar to PAPER-ECON in purpose, but is designed to be used after more site-specific information is available, such as penstock length, tailrace length, transmission line length, and dam repair requirements, so as to better estimate the site economics.

The input and output parameters for PAPER-ECON and HYDRO-ECON are shown in Table 5-7.

Some additional work has been done by others, such as Hydrocomp Inc. of Walnut Creek, California, on the evaluation of proposed hydropower sites. These programs were first developed in FORTRAN for mainframe computers but have since been downloaded onto microcomputers (Fritz, 1984). The principal goal is to develop flow duration curves and maximum and minimum flows from precipitation, evapotranspiration, and topographical data.

There continues to be a great deal of interest in using these techniques to evaluate hydropower resources in developing countries, and the introduction of microcomputers makes this easily possible.

Table 5-7 HCR Program Input and Output

Input	Default Values PAPER-ECON	HYDRO-ECON
Title	N/A	N/A
Date	07/01/78*	07/01/78
Your Name	N/A	-
City	N/A	-
State	N/A*	-
Construction Date	N/A	-
Average Head	O*	O*
Average Flow	O*	O
Percent O&M	2*	3%
Discount Rate	9% & 16%	9%
Revenue Escalation Rate	8%	8%
Expense Escalation Rate	6%	6%
Economic Life	30 years	30 years
Escalation Life	10 years	-
Plant Factor	50%	-
Plant Capacity	-	O*
Dependable Capacity	-	N/A
Energy Value	-	O*
Tailrace Length	-	O*
Penstock Length	-	O*
Transmission Line Length	-	O*
Transmission Voltage	-	13.8 kV*
Earth Volume for Dam Repair	-	O*
Concrete Volume for Dam Repair	-	O*
Road Construction Length	-	O*
Contingencies	-	15%
Indirects	-	25%
Number of Generating Units	-	1

*Required input

Table 5-7 HCR Program Input and Output (continued)

Output		
Plant Capacity	X	-
Annual Power Generation	X	-
Investment Costs	X	X
Overhead & Management Costs	X	X
Average Annual Revenue	X	X
Benefit/Cost (B/C) Ratio	X	X
Internal Rate of Return	X	X
Graph of B/C vs Discount Rate	X	X
Plant Factor	-	X
Present Value	-	X
Cash Flow Rates	-	X

Broadus, C.D., Hydropower Computerized Reconnaissance Package, U.S. Dept. of Energy, Idaho Operations Office, April 1981.

SELECTED EXAMPLES

Energy Planning, Morocco

The origins of microcomputer application for energy sector investment analysis in Morocco began at the World Bank in early 1981. The government of Morocco requested capital for a broad package of energy sector projects. At the World Bank, there was concern that the proposal represented a diffusion of investment capital spread too widely over the energy sector and that a more limited investment portfolio would be appropriate. A meeting was called to bring together energy sector specialists and finance analysts to discuss the energy situation and investment climate in Morocco. The decision was made to prepare an investment evaluation methodology (Gordon, 1983).

In 1982, an energy sector planning and investment system was developed, funded by USAID and called Morocco Energy Investment Strategy. The microcomputer played a central role this project.

Uncertainty is a large concern in developing country investment analyses, in particular with regard to the capital cost of energy technologies, heat rates, interest, debt service, and fossil fuel prices, as well as to engineering and financial parameters. The proper way to handle such uncertainty is to structure a probabilistic investment analysis. But to carry out this analysis by hand is unrealistic; therefore, the microcomputer was brought in.

The approach to energy sector investment analysis resulted from a series of interviews with senior officials advising the minister. The intent was to discern how, in Morocco, energy sector projects are selected for serious consideration as capital investments and how capital is allocated to various projects. An important benefit of this approach was to enlist the active support of a number of officials responsible for different areas of energy planning. Since this effort also was coincident with the beginnings of the Ministry of Energy and Mining (MEM), the development of a major energy sector investment methodology afforded a unique opportunity for project participants to demonstrate the virtues of a systematic approach with microcomputer support.

By early 1983, the system for energy sector planning and investment analysis was applied in formulating investment plans for the Ministry. The planning system required information from subagencies within and outside of the MEM. In effect, the existence of this planning methodology stimulated detailed engineering, technical, and financial analysis of proposed energy sector projects within the country's energy establishment. Any group not providing information to the system ran the risk that its programs would be misrepresented in analyses, and therefore cooperation resulted.

In the fall of 1983, the investment planning system was in regular use in preparing the five-year plans for the MEM. The system running in FORTRAN on the Apple II with a hard disk was soon rewritten in Pascal. Subsequent interest in Costa Rica and other countries has led to a system available in French, Spanish, and English, and on the IBM-PC as well as the Apple II.

By 1984, the energy sector planning and investment system was a fixed tool of the ministry. Subsequent budget cuts of 40 percent issued to the ministry required modification of the five-year plans. Analysis that would have taken weeks now takes days, and the quality of results is improved as well.

Energy Planning, Sudan

This "bottom up" example of energy sector planning began in the Sudan with the first energy assessment in 1980. Prior to that time, there had been selected assessments in specific areas such as renewable energy projects, the oil pipeline, and petroleum exploration and development. The energy assessment, however, was the first comprehensive look at patterns of supply and consumption, and in particular how energy demands might grow and how potential sources of supply would meet this increased demand.

A microcomputer appeared almost spontaneously in the summer of 1982. The energy assessment staff was engaged in demand forecasting, particularly in the collection and processing of survey information for the industrial, transportation, commercial, and residential sectors, for which little concrete information existed. Processing survey data and other information from ministries was becoming burdensome. But help arrived in the form of an Ohio Scientific computer. Immediately, assessment staff began using it do basic statistics, spreadsheet analysis, reports, and demand projections. To familiarize junior staff with computer work, a course in BASIC was set up through the University of Khartoum. By the end of the year, however, the microcomputer was not working well.

At the same time that the Ohio Scientific equipment had problems, the demand analysis work increased. The staff was inundated with demand forecasts, which could only be done by hand computation. The National Energy Administration (NEA) was able to arrange use of the Ministry of Agriculture North Star microcomputers during evenings and other off-time. Additional time was obtained on the University of Khartoum's mainframe. But at NEA, there was need for dedicated equipment and statistical analysis software. Coincidentally, the Georgia Institute of Technology was involved in renewable energy projects in Sudan, using Osborne portables running Supercalc software. So, partly to be able to swap the hardware and software it needed, the NEA obtained its own Osbornes in February 1983. The equipment remains in good working order, with occasional maintenance.

By July 1983, an IBM-PC was purchased. Before this time, no service of any kind was available, but limited support services became available in Khartoum for the IBM. The NEA project acquired a basic 64K PC with VisiCalc and Wordstar software packages. At the time, most of the staff were already using the Osborne and Supercalc. The IBM-PC was therefore used for numerical analysis. However, there was concern for the quality of reports prepared at NEA for distribution to other agencies. Part-time allocation on the IBM was made for word processing, with the predictable result that quantity, timeliness, and quality of reports improved. Six months later, the IBM had become exclusively a word processor.

In 1983, LOTUS 1-2-3 became available. Since the IBM-PC was used only for word processing, in late 1983 a Hyperion microcomputer with 256K and graphics was purchased. This equipment was subsequently utilized in the industry-by-industry, sector-by-sector energy demand forecasts.

An Arabic language word processor for the IBM-PC became available in January 1984, creating the impetus for adding yet more hardware and software to NEA functions. For English language word processing, Volkswriter, which had spelling and grammar checkers, was used. The quality of reports from those not entirely fluent in written English vastly improved. A Hewlett-Packard plotter has been added for presentation of quality graphs. New transformer-stabilizer equipment was also provided for improved line isolation and power outage backup. An air conditioned computer room was also made available.

The NEA subsequently became interested in management information systems for the energy sector. The intent was to provide a system not only capable of the usual energy demand forecasting and other planning activities, but also able to provide more detailed financial and petroleum flow information for day-to-day management. To implement this information system will require an IBM-XT. An IBM-XT, purchased under a grant from the Chevron Oil Company, was already in use in Sudan at the General Petroleum Corporation, a government controlled corporation.

The story of microcomputer evolution at the National Energy Administration in Khartoum sounds surprisingly like the evolution of microcomputer use in developed country organizations. Their adoption is a matter of convenience, easy use, low cost, and finally indispensability, leading users to wonder how they previously managed their work.

Energy Planning, Sri Lanka

The use of microcomputers for energy planning evolved in quite a different way in Sri Lanka, being driven from the outset by policy considerations and the institutional framework. To strengthen energy management, a series of task forces was established to provide policy coordination and focus for various planning functions. The Energy Planning and Policy Analysis (EPPAN) Task Force was charged with the development of a national energy strategy, and the microcomputer based energy planning models were designed with the particular analytical needs of this task force in mind.

The energy modelling project was based on the concept of Integrated National Energy Planning (INEP) developed in the late 1970s (Munasinghe, 1980a and 1980b). The INEP procedure was initially implemented on a preliminary basis in 1982, using manual methods (Munasinghe and Schramm, 1983). From this learning process grew the necessary confidence to embark on the more sophisticated hierarchical microcomputer modelling framework.

The practical requirements of the policy focus dictated the design of the modelling framework. In place of an all-embracing single model, a hierarchical structure was adopted, in which individual models at subsectoral, sectoral, or macroeconomic levels could be exercised alone, or linked with each other. The individual models were implemented in stages; the first software to analyze supply-demand balances and their foreign exchange implications was operational within the first few weeks of the project. Data validation and staff training were initiated at the very outset.

As shown in Figure 5-3, two subsectoral-level models were developed. The BMASS fuelwood model is designed to analyze the impact of both fuelwood planting and efficient cookstove programs on the accelerating deforestation. A linear programming model (REFLP) of the refinery and petroleum products sector, involving about 60 rows and 180 variables, was also built. These two models, together with the electric capacity planning models such as WASP, provide key inputs to the next hierarchical level of modelling.

The RESGEN physical energy balance model is based on the Reference Energy System network; it provides a complete progression from energy demand, through distribution, transmission, and conversion, to energy supply. THe EFAM model simulates the income statement, balance sheet, and sources and uses of funds for various entities, including the electric utility company, oil and gas companies, the government, central bank, and overseas lending institutions. The model keeps track of both domestic and foreign currency flows, and is particularly suited to analysis of pricing policy issues.

The model at the top of the hierarchy, ENMAC, is a multi-sector simulation model that is designed to capture the chief interactions between the energy sector and other key macroeconomic sectors. It has a transactions matrix involving 10 sectors, including three energy producing sectors, crude oil imports, and six energy consuming sectors. The model produces a set of national accounts, balance of payments, and equilibrium wages and prices. Production functions are used to generate sectoral outputs, as the model forecasts year by year into the future.

By late 1983, when the project started, MS-DOS had already assumed the status of a de facto standard for 16-bit microcomputers, and the hardware specifications posed few problems. However, given the importance of in-country maintenance access, the decision was made to purchase the machines in Sri Lanka. Since at the time the decision to use the IBM-PC had not been officially announced in Colombo, the WANG-PC and the TANDY 2000 were chosen. Both have proved reliable and compatible with each other and with IBM, and in fact in some respects are superior to the PC because they are true 16-bit machines based on the 8086 chip; they are therefore considerably faster than the 8088-based IBM-PC and XT. The WANG is installed with a hard disk, and the TANDY with high resolution color graphics capability.

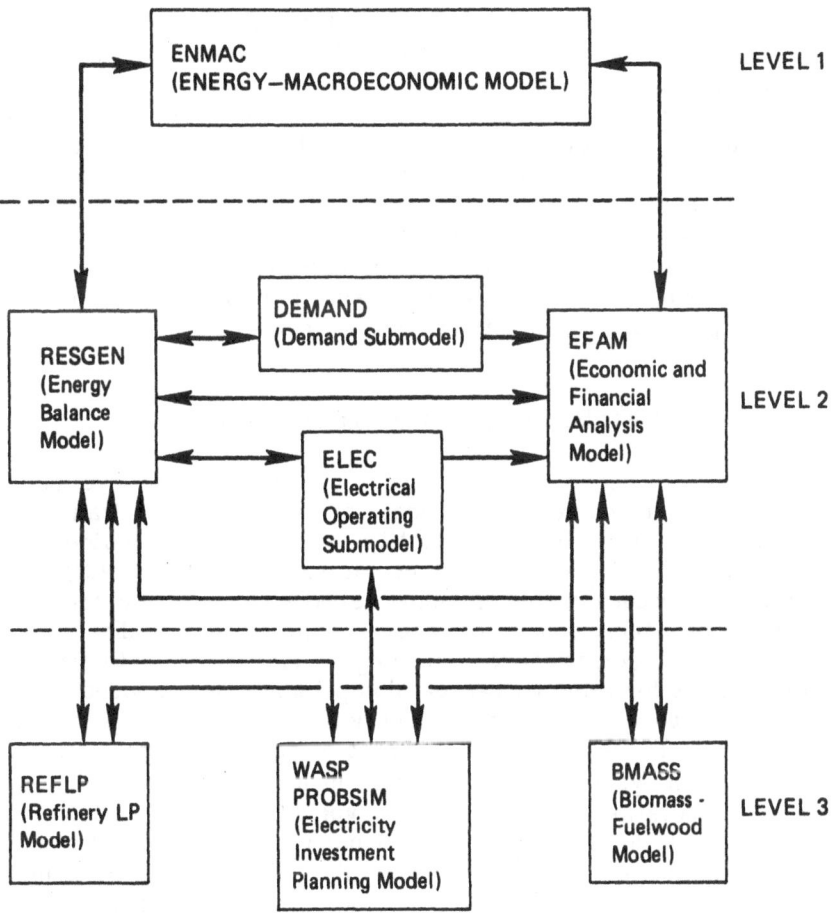

Figure 5-3   Hierarchical microcomputer modelling framework for
integrated national energy planning

The software itself is quite diverse. LOTUS 1-2-3 is used for many routine data analyses and is the basis for file manipultation. The RESGEN and EFAM models are written in FORTRAN enhanced by assembly language routines; they were adapted for use in Sri Lanka. The refinery model is based on a commercially available LP package.

The experience in Sri Lanka points to several ingredients for the successful use of microcomputers. First, policy considerations should dictate the choice and design of the model, not the other way around; all too often, models are developed without any clear application in mind. Second, whereas microcomputers can facilitate analysis, they are no substitute for the solid educational and professional experience of the analyst; a key factor in the successful use of models in Sri Lanka is the background and training of the staff involved. Third, adequate attention must be given to maintenance issues and the operating environment at the point of hardware purchase. There is little benefit in buying offshore at discount prices if subsequent maintenance cannot be assured.

Distribution System Design, Sri Lanka

For a number of years, power system planners have been familiar with computer applications for generation and transmission studies. These include load flow studies indicating line and transformer flows and bus voltages for various system configurations, short circuit studies providing network currents and voltages under various fault conditions, and dynamic system studies simulating system performance following a disturbance. Computer based studies for generation-expansion planning also included some interaction with such economic considerations such as capital and operating costs. In addition, categories of power system computer based control systems have been used in load dispatch centers to allow the operation of complicated networks (Ratnayake, 1984).

In the field of distribution planning, power utilities were content, until recently, to use rough guidelines and rules of thumb developed over the years. Thus, conductor sizes, loading norms, and areas of coverage per distribution transformer had become standardized with practice. In many instances these standards have later been found to be quite uneconomical with respect to modern conditions. The oil price rises of 1973 and 1979 forced power utilities to look for cost saving measures, such as improved system design.

In a typical utility in a developing country, almost half of all losses occur in the distribution system. Further, distribution investment constitutes about a quarter of the capital budget. In the recent past, these factors have drawn the attention of power utilities to the necessity of optimizing loss levels with respect to investment costs. Thus arose the necessity of modelling distribution systems by computer for both the database and associated technical and economic analysis. In many developing countries, it was necessary to change the distribution systems' design criteria from purely technical considerations of line voltage drop to economic considerations of minimizing of system losses and associated costs.

Database Establishment

The relative complexity of establishing an accurate database and the high cost of computers were two causes for delay in using techniques such as load flow studies for distribution systems. However, with the introduction of microcomputers, especially new models with greater memory, new planning activities could be contemplated. A typical database consists mainly of:

- Regional load forecasting
- Determination of aggregate consumer characteristics, such as load variation according to time of day and seasons, and load duration curves
- Representation of the existing system and proposed development lines.

In the following section, microcomputer based techniques for developing the database are explained.

Regional Load Forecasting. Three basic techniques exist for developing system load forecasts:

1. Building up an overall forecast from subforecasts of consumer categories (such as domestic, industrial, commercial)
2. Using trend methods and historical data to yield future expectations.
3. Simulation of land-use data and expected growth patterns.

In building up a forecast from consumer classes, the initial load pattern is assumed to remain constant and the magnitude of each class is increased by an expected growth factor for each year. A computer program can compute the 24-hour load curve for each year for each class of consumer, producing a cumulative load curve. Forecasts made for each area are integrated to give the demand expected by larger geographic divisions such as a main feeder or a grid substation. In addition, historical data fed into the computer can be used to develop a best fit growth curve. These

programs should also be capable of accepting the analyst's estimates of load saturation for each small area as checkpoints for
target years.

With the method described above, land use models can also be
developed and programs written to translate the expected land use
data into electricity demand. For this purpose, city planning maps
may be used to identify the utilization type of areas or "zones."
Based on these maps, the expected population and commercial and
industrial floor area by various subcategories could be used to
forecast specific electricity consumption. Special programs can
identify patterns of land use, such as the tendency for commercial
development to occur near major street intersections and the
proximity effects of city centers.

Determination of Consumer and Line Loading Characteristics.
Load forecasting is, to a large extent, dependent on the assumptions made for intensity of loading by various consumer types and
the nature of their daily or seasonal load profiles. Data analysis
available with computerization could effectively be used to obtain
correlation and trend factors. The monthly consumption data are
in the computer if billing is computerized; hence their analysis only
requires a slight extension of the program to store, sum, categorize, and compare data.

This type of analysis can yield:

● Breakdown of consumption according to customer type for
each selected sub-area
● Historical load growth pattern for each consumer type
● Identification of seasonal variations
● Analysis of the change of consumption trends among
existing consumers as well as the load increases because
of new consumers
● Aggregation of consumption by transformer station or
distribution feeder.

In addition to information on consumption patterns and trends,
the relationship between the power demand (kVA, during system
peak) and energy consumption (kWh) must be determined. If time-
of-day metering is used, these relationships could be captured
without difficulty. If not, however, feeder and substation current
measurements taken during the billing period could be combined to
develop the required data for the particular consumer mix.

A more sophisticated and comprehensive method for obtaining
the information above is via electronic meter to microprocessor
based demand recorders. These units are used along with portable
readers or even telephone line communication systems. Telephone
line communication systems can be used for both comprehensive

load surveys and load control applications. However, it may be prudent to keep in mind that microcomputer based systems providing complete time-of-use consumption data and full integration of metering, communications, and data processing will eventually become commonplace in electric utilities in developing countries. Representation of the Distribution System. To study the performance of a power system, it is necessary to represent the system by way of line sections, with electrical characteristics such as loads nodes. This information could be fed into the database after measuring line lengths and calculating electrical characteristics. However, the most convenient way of establishing the system database is via a digitizer. The map of the power network is overlayed on a digitizer, and a cursor or pen is used to transfer the x-y coordinates to the computer. Complete information transfer is made possible by a "menu" used with the digitizer. Any section of the power network can be selected and the required analyses undertaken. Similarly, any proposed grid extension proposal can be included via the digitizer, and its impact investigated. More will be said about these techniques in the following section.

Distribution System Studies

Microcomputer based distribution system studies are also important for system operation. The most useful information for distribution systems is load flow, which provides information on current flows in individual sections, attendant line losses, and the voltage profile at the system nodes.

These studies are most important during system peak hours, when the maximum load occurs. However, studies of the system during other times (such as sub peaks and base load times) are also important, particularly in computing the flows and energy losses. Studies are also required for such contingency situations as the loss of a feeder or changed switching configurations.

Other studies that could be conducted on power systems using microcomputers are:

- Fault Studies—to give fault currents under line to ground or three-phase fault conditions
- Capacitor and Regulator Optimization—to determine preferred locations of fixed and switched capacitors and regulators
- Reliability Analysis—to calculate reliability indices for various system operation conditions
- Protection Coordination—to coordinate setting of circuit breakers, relays, reclosers, and fuses.

System Studies and Technoeconomic Considerations

System studies could be used to optimize operating conditions on an existing distribution system. However, the main purpose of such analysis is to establish optimum development plans for future expansion. This involves the study of alternative development proposals, obtaining the operating characteristics of each system, and subjecting the alternatives to economic scrutiny. Technical criteria that need to be satisfied for each development proposal are voltage limits and current carrying capacities. Both values must be acceptable for normal operating conditions and for exigency conditions that the system is expected to bear. Proposals that are technically acceptable can then be evaluated economically. The preferred solution would be the one with the least present value cost considering these factors:

● Investment or capital costs
● Lost revenue due to losses or theft
● Suitable expected value of reliability differences between
  alternatives.

Distribution system planners should keep in mind that optimizing individual feeders is not the same as optimizing the system as a whole. Hence, full use of the digitizing facilities should be made to compare alternatives that could alter existing feeder networks.

The microcomputer and associated hardware and software (plotters, digitizers, and statistical and optimization packages) all come into play in planning improved distribution systems. Although there is software available, most of it is proprietary, available through a variety of consulting organizations.

Training in Energy Planning, Africa

The microcomputer was a central and integral part of recent seminars titled "Designing and Implementing a National Energy Conservation Program," held in Nairobi, Kenya, and Dakar, Senegal. The IBM-PCs provided relief from the slowness and tedium of "number crunching," so the participants could better focus on the issues. The use of the microcomputer in this instance also provided extensive "hands on" analysis of energy conservation opportunities with real data (Gillings, 1984).

These Africa seminars were a joint effort of Acres International in Canada and Lavalin in France. By 1983, a base of software and hardware experience could be readily applied to the design of short courses. The IBM-PC was adopted by Acres

International as standard equipment and found application in Nepal and elsewhere for power system planning, economic evaluation, and macroeconomic forecasting. Also, by this time, dealer networks had extended sufficiently into Africa to support the growing use of microcomputer equipment. For example, dealers for the IBM-PC were in Senegal, Nigeria, Ivory Coast, Cameroon, and Kenya. Apple and Hewlett-Packard dealers were in Nairobi and Dakar. International Computers Ltd. was in Nairobi. With this expanding network, the time was ripe for training that incorporated substantial dependence on the microcomputer.

To simulate the experience of national energy planning in a two-week seminar was difficult. The dual objectives of the seminars were (1) to have participants understand the fundamental concepts of commercial and economic feasibility and relate these to realistic projects to improve the country's energy supply-demand balance, and (2) to have participants do sufficient calculations to appreciate the strengths of such analysis and the practical difficulties in carrying out energy conservation analysis. However, the final design of a national energy conservation program required processing and analysis of an enormous range of data and information, followed by specific project formulation.

Within the brief time allotted, the intent was to take a comprehensive look at the national energy situation in a developing country, examine a broad range of conservation opportunities in all sectors—transport, industry, commerce, residential—and make a realistic assessment of the technical, commercial, and economic feasibility of conservation options that might be included in a national plan. Admittedly, this was an ambitious undertaking, and to conceive of such a range of activities without microcomputer support would have been impossible. The microcomputer played a central role in two aspects of these seminars. First, the IBM-PC was used extensively in the compilation of data to prepare the case study, and second, it was used extensively for quick number-crunching.

The seminars included (1) a series of lectures on the fundamentals of economic evaluation of energy conservation opportunities, (2) the use of Terrania (a fictitious developing country based loosely on Kenya) to provide realistic and comprehensive data, and (3) the use of Lotus 1-2-3 on the IBM-PC as the computational tool. The thirty participants were broken into four groups with access to the two IBM PCs brought to Nairobi and Dakar for the seminars. The use of Lotus 1-2-3 afforded the ability to perform net present value and cost-benefit analysis of many alternatives quickly, to integrate these into an implementation scenario, and to assess the impact of the proposed program on the projected energy supply-demand balance of the nation.

Such seminars or workshops are an ideal mechanism for bringing energy conservation technologies to the developing countries; the use of the microcomputer as a tool for analysis and data management is vital to their success. These types of analyses cannot be done without a computer. The availability of commercial software, such as spreadsheets and data management packages, gives the analyst all the tools required.

Microcomputer Based Energy Data Acquisition Systems, The Philippines

To maximize the use of new software, data must be available to the microcomputer. It is usually entered via keyboard or disk, and for certain manufacturing and scientific applications, this approach is slow and error prone. Data coming from instruments measuring voltage, temperature, and pressure, for example, are necessary for process control research. An alternative approach to keyboard entry is through an automatic data acquisition system using direct measurement, storage, and processing of data gathered from measuring instruments (Alcala, 1984).

Previous data acquisition systems were based solely on minicomputer or even mainframe computers and were thus beyond the reach of most industries and institutions. These systems used centralized discrete analog signal conditioners and analog to digital converters for measuring the electrical signal coming from sensors. Applications were thus limited to large process control systems and scientific research and development projects.

The development in microcomputers has made data acquisition easier and has lowered costs. The signal is read from each sensor and is partially processed by a dedicated microcomputer based data acquisition system before it is sent to a central processor using local-area-network technology. Often, a microprocessor based measuring instrument fitted with a suitable digital communications port qualifies as a distributed data acquisition system component. Several new digital multimeters with IEEE-488 or RS-232C interface are examples, and their availability has made automatic data acquisition practical for small users.

Examples of an Application

An automatic data acquisition system set up for a thermal processing plant such as a rotating kiln, for example, requires a means for continuously measuring the weight of raw material or feed rate, temperature, and pressure, as well as other mechanical and electrical data. Transducers, such as load cells and thermo-couples, are used to convert mechnical force and temperature to an equivalent proportional electrical voltage. The data acquisition system converts this information into digital form, storing it for subsequent processing and report generation. Such a design relies heavily on single chip microprocessors for partial data processing at each remote location. This approach results in a marked reduction in wiring, higher reliability, and a release of overhead processing by the central computer.

Software

Most data acquisition systems are sold as complete systems to end users and configured for particular applications. Usually, the only option available is a proprietary operating system and data acquisition software, and these are often incompatible with some of the popular operating systems. Also, there are a number of sophisticated instruments that are microprocessor-based but do not have the capability of running any applications programs except those supplied by the manufacturer.

However, there are a number of excellent software programs written in CP/M and MS-DOS that can be used with data acquisition systems. The combination of dBase III for storing and arranging data, LOTUS 1-2-3 for spreadsheet analysis and graphics, and Turbo Pascal for computations is appropriate for data acquisition systems. Files can easily be interfaced between each of these packages using several utilities that enable different application software to load data from a variety of programs. The effort required to set up a CP/M or MS-DOS compatible data acquisition system is justified because of lower commercial software cost.

A data acquisition system has an immediate payoff by providing rapid analysis and a well-organized database. The use of microcomputers for this purpose is still recent. It can be expected that in the near future, software packages will be available specifically for data acquisition.

Building Energy Use and Control, India

Following the sharp increase in oil prices in 1973, attention was drawn to the problems of improving the end-use efficiency of energy for buildings. In the developed countries in temperate and cold climates, as much as 30 percent of the national energy use was directed to heating and cooling buildings. This sector was recognized to be energy inefficient, and research on passive solar heating and cooling techniques for buildings began to attract serious attention from the mid-1970s (Gadgill, 1984).

In developing countries, cities continue to grow. Tall commercial buildings will be part of that growth, consuming up to 30 percent of a country's electrical energy production, primarily for air conditioning. This represents a vast opportunity for conservation and energy efficient design carried out through microcomputer based techniques.

The electrical energy consumption in a large office building can total several megawatts. This load often requires peak power from supplementary oil fired generators. The design philosophy for commercial buildings is beginning to change as oil prices continue to increase. For example, passive solar architectural design, which has evolved over hundreds of years, particularly in North Africa, is gaining popularity.

With the addition of new design tools, the microcomputer and engineering software that has evolved over the years by trial and error can now be programmed or simulated mathematically both in the field and the design office. A large variety of software for building design is currently available. A recent survey of design tools for passive and hybrid low energy buildings was undertaken as part of the International Energy Agency Solar Heating and Cooling R&D Program. Completed in late 1982, the survey revealed that there were some 59 design tools made for mainframe computers and 94 for microcomputers (Ahmed and Ritterman, 1984). With increased capacity available in the new microcomputers, one can assume that there are additional design tools available today.

The need for research in building energy use can be justified by examining urban growth in India. It should be noted that the pace of urbanization is more rapid than population increases in most developing countries, and India is no exception. For example, the energy consumption in commercial buildings in India for the years 1975 through 1985 is shown in Table 5-8. The current annual growth rate of energy consumption in Indian commercial buildings is about 8.7 percent, much larger than the annual population growth of about 2.2 percent.

Despite this rapidly increasing energy consumption, Indian schools of architecture did not offer courses on energy management until recently. Furthermore, computational facilities for architectural students and faculty were nonexistent. In addition, architectural firms could neither afford access to mainframe computers nor had training in the use of computers to simulate energy performance of buildings that they designed.

Because microcomputers are becoming affordable and accessible, increased application of building energy simulation programs can now be considered. Until 1983, no computer program to simulate energy consumption in buildings was operational in

India. However, by 1984, some mainframe based programs were introduced at the Indian Institute of Technology in New Delhi.

Table 5-8   Electricity Consumption of Commercial Buildings in India (Units of 10^{12} Watt Hours)

Years	Energy Consumption
1970-71	2.71
1975-76	4.10
1977-78	4.43
1980-81	5.70
1984-85	7.31 (estimated)

Gadgill, A. "Microcomputer Model of Thermal Performance and Energy Consumption of a Residential Building," paper presented at the Symposium on Microcomputer Applications in Developing Countries, Colombo, Sri Lanka, 1984.

TWOZONE

TWOZONE was one of the several computer programs developed in the United States during the late 1970s for modelling the thermal performance of buildings to determine energy consumption and identify potential cost-effective conservation strategies in design and management. The background and mathematical equations on which TWOZONE is based were first published in 1977 (Gadgill et al., 1978). TWOZONE, as the name implies, is a building model with two thermal zones for determining energy use and comfort conditions in residential buildings, and for analyzing the impact of various energy conservation strategies. In terms of model sophistication, TWOZONE was soon superseded by large complex computer programs such as DOE-2 (Lokamanhekim et al., 1979), and BLAST, which were developed on mainframe computers widely available to U.S. institutions. Today, most of these programs, including TWOZONE, are available on CP/M or MS-DOS based microcomputers.

A feature of TWOZONE is its straightforward computational structure. It lends itself well to modifications and alterations by users. The user accessibility of the FORTRAN code in TWOZONE encourages experimenting with the building model itself. Since nontechnical users often tend to have a misplaced faith in computer generated results, this capability makes it less mysterious than if the program remains in the form of a black box.

Secondly, TWOZONE is flexible as to the configuration of the building being simulated. This is particularly important in India, which has a wide variety of styles of construction and materials in residential buildings. TWOZONE allows the user to define the walls and roofs layer-by-layer, making itself adaptable to a wide range of construction types. It is a simple matter for a user of TWOZONE to modify the program and incorporate known correlations (e.g., for infiltration, ground conduction, or wind ventilation). User-written subroutines simulating any new active or passive solar system can also be integrated into the building model. Predictions from TWOZONE have been favorably compared to predictions for identical buildings with DOE-2.

The Lawrence Berkeley Laboratory, where TWOZONE was developed, has published a users' manual in response to queries from several users in the United States and abroad. The microcomputerized version of TWOZONE has retained all the features of the original version, and thus the manual is still applicable

Program Description and Use. Input for TWOZONE consists of the following:

- Heat transfer functions for walls and roof, building description schedule for internal heat loads, thermostat settings, description of windows (areas, orientation and number of panes).

- Air conditioner or evaporative cooler capacity and power consumption, and hourly values of weather data for the entire simulation period (8,760 hours for a one-year simulation). TWOZONE was originally designed to read weather data from weather tapes, which are available for most American cities. It also can read data from weather tapes prepared for DOE-2. However, for most regions of the developing world, weather tapes are hard to obtain, and TWOZONE has facilities for constructing approximate hourly weather data internally when supplied with some indicators of the local weather.

- If desired, TWOZONE can calculate pay-back periods for various alternative strategies based on additional energy and economic data that the user can supply. It also supports a simple printer-oriented graphics package for displaying the thermal simulation results.

For each hour, the program computes the transient one-dimensional heat transfer through each wall and the roof, solar gains based on the cloud cover and window reflections computed from altitude and azimuth of the sun, internal heat loads based on schedules, air conditioner or evaporative cooler performance based on thermostats and psychrometry, and infiltration gains and losses based on the wind speed. As described, the building is modelled as two zones in thermal communication with each other by natural convection. Heat balances for each zone are calculated hourly based on the average indoor temperature, thermostat setting, and cooling or heating from appliances.

If the outdoor temperature is lower than the indoor temperature, then the model considers the option of simply opening operable windows rather than switching on cooling equipment. Other strategies (for example, using insulating reflective curtains on windows during the day) can be easily programmed.

Hourly weather data are required by the model to calculate solar radiation from observed cloud cover and reflection from window pane surfaces. The following data are needed:

- Outside dry-bulb and wet-bulb temperatures
- Cloud cover and type
- Wind speed
- Dew-point, humidity ratio, enthalpy, air density, and barometric pressure.

Typical program output consists of:

- Hourly air conditioner or evaporative cooler load for the first four days of each month
- An additional printer plot of hourly energy use
- Hourly cooling load and heating load distributions over the other days of each month
- Summary output for the entire run period including apportioned heat losses and gains and apportioned cooling and heating loads from windows, walls, roof, floor, and infiltration.

Microcomputerization. TWOZONE originally required 125K of memory, making it impossible to run on early 8-bit microcomputers. The program, therefore, was divided into segments, and each segment executed sequentially. All segments except the first receive data from their predecessors by an intermediate file and all except the last write a file for their successors. There are a total of five segments, of which only two are essential for executing the program. The remaining three segments are useful for exercising the various optional features available with TWOZONE such as graphics, economic analysis, and detailed air conditioning calculations.

The first segment is used simply for reading the input and creating a sequential file of data that is used subsequently. In the second and third segments, calculations are carried out for the entire run period (8,760 hours for a one-year simulation). Each consists of two parts: one devoted to calculation of heating gains and losses from the external environment (including internal loads) and the second to determine the rate of heat extraction from the internal environment by the air conditioning or evaporative cooler. These segments differ from each other only in terms of the details of computations.

In segment 2, the heating gains and losses from the external environment are calculated for each hour. The results are then passed to the cooling component where, depending on its capacity and thermostat settings, a new indoor temperature is computed. Heat gains for the next hour are also calculated. During its execution, segment 2 writes an intermediate file that is used as input by segment 3.

In segment 3, in place of computation of heat gain and loss, data from the intermediate file written in segment 2 are used to drive the cooling appliances, whose engineering performance is simulated. This simulation includes computations such as the amount of water consumed by the evaporative cooler or the amount of condensation on the cooling coils in the air conditioner. Segment 3 summarizes the net energy consumption, apportions cooling and heating loads over the run period to windows, walls, roof, floor, and infiltration, and provides the hourly load profile averaged over the run period for electricity demand.

Segment 4 performs an economic analysis. Using data provided by the user on the "base case" energy consumption, economic data, and costs of various energy conservation measures and strategies, segment 4 calculates pay-back periods given such factors as inflation rates and depreciation rates.

Segment 5 provides the graphics output to a printer. This is considered extremely helpful in communicating model results to nontechnical users.

Since the entire hourly simulation is done twice, once in segment 2 and then again in segment 3, the execution time is long. A two-month simulation without economic analysis or graphics required two hours on a system with a clock speed of 4 MHz. Microcomputers have very low operating costs; therefore, programs that run for several minutes or even hours are acceptable to those who are not pressed for time and have no ready access to larger and faster machines.

Conclusion

Mainframe computational facilities are not available to most educational institutions dealing with architecture in the developing countries. Microcomputer programs that can be easily modified by new users based on simplified algorithms have great potential for training the current generation of students in architecture to quantify their concepts of energy conservation. The microcomputerized version of TWOZONE is expected to be useful in filling this gap.

Fuelwood Model: BIOCUT, Liberia

BIOCUT was developed to facilitate the systematic investigation of a wide range of potential wood energy plantation applications and to provide reasonable estimates of net returns and minimum required output prices. The model was used to conduct preliminary economic investigations of wood energy supply for rural electric power plants in Liberia. To take advantage of the flexibility and convenience of microcomputers, a mainframe based model, FIRSTCUT, the precursor, was modified. The microcomputer based model, BIOCUT, retains many of the essential features of the mainframe model while at the same time being transferrable and field analysis oriented, a major advantage in conducting on-site economic evaluations (Perlack et al., 1984).

The BIOCUT model was developed in full recognition that the evaluation of wood energy plantations should be viewed from a general economic perspective. BIOCUT is therefore intended for preliminary economic assessments in which the principal analytical objectives are: (1) to identify the most promising plantation designs (for example, rotation ages, spacings) and silvicultural management alternatives, (2) to indicate the nature of potential tradeoffs in system design and operation, and (3) to facilitate extended sensitivity and risk analyses.

In BIOCUT, the biomass plantation is modelled on the basis of a large number of user-specified activities. These activities can be tracked separately to indicate their impact on major production cost components. The description of each activity is general enough to apply to a wide range of plantation designs. The user-specified activities in BIOCUT are grouped into five general categories: (1) plantation scenario activities, (2) economic parameters, (3) plantation establishment activities, (4) operation and maintenance activities, and (5) harvesting, processing, and transportation.

The model can be used to analyze the economics of woody biomass production in two modes. The first mode assumes that the plantation is designed to provide a continuous yearly flow of products beginning with the first harvest. The model accomplishes this by planting only a fraction of the total plantation in each of the established years, depending upon the user-specified rotation age. For example, if a four-year rotation age is specified, then one-fourth of the plantation is planted in each of the first four years. The second production mode does not assume a continuous synchronized forest, but establishes the entire plantation at the outset.

The BIOCUT model calculates four summary economic evaluation criteria: (1) the net present value, (2) the discounted average cost, (3) the internal rate of return, and (4) the cost-benefit ratio. The maximum net present value is used to define economic efficiency for the biomass plantation. Although there are no specific optimization routines in the model, the user may readily optimize certain aspects (for example, determining an optimal rotation age) of the plantation design by using the model in an iterative context. The internal rate of return and cost-benefit ratio are alternative evaluation criteria to the net present value. The discounted average cost is the minimum selling price required for the energy plantation to be viable over its economic life.

In economic evaluations of wood energy plantations, there is clearly a high degree of uncertainty surrounding important biological and economic relationships. The BIOCUT model is therefore specifically structured to expedite sensitivity analysis, permitting the detailed examination of the effect of changes in user-specified activities on economic criteria. If the user is able to specify or describe the range of uncertainty for model cost activities and revenue, then the model can be used to carry out risk analysis as well.

Actual experience in Liberia provided the incentive to reprogram and modify the mainframe based computer model, FIRSTCUT, for the microcomputer. Wood energy plantations could become a necessary long-term sustainable resource in Liberia as existing wood energy supplies are depleted and as demands for wood energy feedstocks increase.

One of the first questions to resolve in analyzing the economics of wood energy plantations is to determine rotation age. As previously discussed, this can be accomplished by using the model in an interactive context; that is, the user can vary the rotation age specified in the model and select the one that maximizes the net present value. In BIOCUT this is performed by changing two options, and other activities that may be involved are automatically adjusted by the model. For example, spacing determines to some extent the likely range of optimal rotation ages. Results indicated that the five-year rotation age provided the maximum net present value at all possible prices and real discount rates. In a more sensitive example, one could expect that the optimal rotation age is shortened by increases in price and by high real discount rates, and lengthened with decreases in selling price using low real discount rates.

Because of the flexibility of the model, extended sensitivity analyses are easily performed. Productivity assumptions and input costs were systematically varied for the five-year rotation age plantation. The base case assumes a selling price of $40 per dry metric ton and a real discount rate of 12 percent. The plantation is particularly sensitive to changes in selling price and yield. Transportation and harvesting are the next two most sensitive activities. But planting and management are not particularly important determinants in the economic availability of the plantation.

In summary, the model was designed for preliminary energy assessments in which the basic economic question is whether and under what conditions wood energy plantations can provide feed-stocks at acceptable costs. BIOCUT was not designed, nor is it suitable, for detailed financial analyses. In general, such analyses are only appropriate after reliable information exists on the most efficient plantation design, on the costs and returns to factors of production, and on the probability distribution of uncertain or inherently variable cost and output parameters. The BIOCUT modelling effort involved difficult choices between flexibility of application and incorporation of detail. Flexibility and ease of use were considered the more important attributes of a microcomputer model. The model was designed so that all input requirements are the responsibility of the user. The flexibility of BIOCUT thus places considerable responsibility on the user, since he must ensure that inputs are reasonable and consistent in the light of available data.

6

Applications
in Municipal Management

INTRODUCTION

The rapid development of microcomputer technology in the last several years could have tremendous implications for information management in the secondary cities of developing countries. Computerization has been slow to reach these cities in Africa, for example, because of obstacles discussed in previous chapters--high initial cost, demanding personnel requirements, and lack of maintenance support.

While the use of microcomputers in developing countries is expanding rapidly, their application to municipal management is moving slowly. There seems to be no coordinated program for introducing the technology, but a pattern is emerging. As mentioned, the technology is appearing first in development projects or feasibility studies, tending to spread through central planning ministries and other central government agencies. In some countries, a second thrust can be seen through the private sector, where vendors of computer equipment have been able to introduce microcomputers through their representatives.

But microcomputer technology has yet to reach secondary city government in any meaningful way in the developing world. However, one can project a significant role in the future for such computers in municipal management even in relatively small towns. Four important roles are emerging; in order of importance they are:

- Records management
- Financial analysis
- Geographical information management
- Staff training.

The role of the microcomputer in records management heads the list because it has an immediate financial impact on municipal

155

government in the areas of tax records (including property reg-istry), utility billing, and municipal personnel records. Improved property tax records systems and utility services billing systems can enhance collection rates, and in many cases they pay for themselves in a short period of time. The specific design of a tax record system prototype is discussed later in this chapter.

Microcomputers have immediate application in municipal financial analysis, both in analyzing current financial positions and in evaluating potential capital investments. For example, the Research Triangle Institute (RTI) has used microcomputers to analyze the yield of various revenue sources across a number of cities in the same country. Comparisons are made that identify the particular city, or group of cities, where revenues have the greatest potential for increase (McCullough, 1985).

In evaluating capital investments, microcomputers can be used to carry out financial analyses expeditiously on different alterna-tives and assumptions (for example, differing interest or inflation rates). This allows city officials to assess the impacts of invest-ments, comparing "worst case" and "best case" scenarios, to see the magnitude of risk they are incurring. In Brazil, for example, a model is being developed that projects cash flow requirements and revenue raising potential through the life of an investment, pin-pointing realistically when the local government can best con-tribute local matching funds.

Municipalities provide a variety of basic services, such as water, sanitation, electricity, and transportation. To plan for their use, a geographical information database showing underground pipes and electric power lines (both subsurface and above ground), must be developed. Microcomputer based geographical information systems (GIS) can now be used for planning purposes. These sys-tems are becoming available at low cost, allowing municipalities to eliminate bulky maps and files.

In the municipal area, the training uses of microcomputers are just now beginning to be explored. Two important categories of use have already emerged: (1) programmed instruction, and (2) training exercises in financial analysis. Microcomputers are used exten-sively in financial management training courses at RTI because they permit participants to carry out different types of financial analyses quickly. Although programmed instruction has not yet been applied to training municipal staff, it holds great promise for providing targeted skills training, which can be carried out in the participant's own office. For example, programs can be prepared for training newly hired staff in established accounting procedures, or for upgrading skills of existing staff.

APPLICATIONS REVIEW

Microcomputer Tax Records System

From the foregoing discussion, it is clear that microcomputers can play an important role in many aspects of municipal management. A good starting point is to begin with local property tax records systems, since the payoff appears greatest in the near term.

Evidence from several studies shows that property tax collection rates are generally low in a number of countries. However, collection rates can be greatly increased by improving the speed and accuracy of tax billing and by early pursuit of nonpayers.

In many cities (and particularly secondary cities) the tax records are so chaotic that some local property owners may never receive tax bills, and nonpayers are seldom identified or dunned for payment.

Furthermore, property valuation is often inconsistent, leading to taxpayer resentment and tax avoidance. Finally, because of the time required to process property valuations and issue tax notices, years may pass between building completion and collection of the tax. In the interim, the city loses needed revenue.

Computerization of property tax records can alleviate these problems. A computerized system operates very much like an existing manual tax records system, but with several important differences. Transactions can be carried out at electronic speed. For example, where transferring information from the property registry to the annual tax role may take a total of five to ten minutes per property when done by hand, the computer can carry out that same activity for several thousand properties in the same amount of time. A computerized system can also issue tax notices at high speed and track payment records instantaneously. Second notices can be sent quickly and lists of delinquent taxpayers can be compiled rapidly.

The labor-saving aspect of the computerized system is important because it frees tax office staff from clerical work and allows them to pursue tax collection more aggressively. The lack of aggressiveness on the part of tax collection staff is a major factor in low collection rates.

A computerized system can also be used to standardize assessment practices. Information can be recorded on property characteristics that allows the tax office to check on the consistency of property valuation. Undervalued properties can be readily identified. Indeed, such a system can be used to carry out the actual valuation of property if objective rating systems are used to assign value based on characteristics. Most tax systems do not permit this yet, but it is a trend in property tax assessment.

RTI has developed a microcomputer program for property tax records management designed specifically for secondary cities in developing countries. The initial application is being carried out in Tunisia with the Ministry of Interior, the Directorate of Local Government Finance. Indeed, the original concept of the system comes from work done in the city of Sousse, Tunisia, in automating the local property tax billing system.

The system was developed with several key objectives in mind. First, it has to manage all the transactions that occur in processing local property taxes. This includes recording valuation information, updating tax rolls, issuing tax liability notices, issuing tax bills, recording payment, identifying nonpayers, and issuing them notices. The system also allows for information on housing characteristics to be recorded so that valuations can be checked for standardization.

The system produces standard reports, such as percentage of collections and amounts collected; furthermore, it permits special reports and analyses to be carried out. For example, if the city is considering applying a betterment levy to recover costs of paving a street, values of all property on that street can be quickly assembled to assess the impact of the levy on the tax bill.

RTI has designed an Automated Property Tax Records System that is a general management information system based on commercially available software from the United States. This system is then customized to individual applications. For example, in Tunisia the system is tailored specifically to tax procedures as specified by Tunisian law.

Despite the need for tailoring the individual applications, the general design of the tax records system is always the same, composed of records of property valuation, tax assessment, tax payment, and mechanisms for issuing notices. In addition, information on property characteristics can also be maintained. The system is shown by the diagram in Figure 6-1.

The program was designed to replicate the existing elements of a typical manually operated tax records system, so that it could be easily learned by local tax office staff. Moreover, most tax systems are rigidly defined by national legislation; therefore, radical departures from existing systems are not possible without legal reforms. It is "driven" by information fed in on valuation of individual properties. Each property has a basic file with additional files linked to it (such as payments, appeals, and property characteristics).

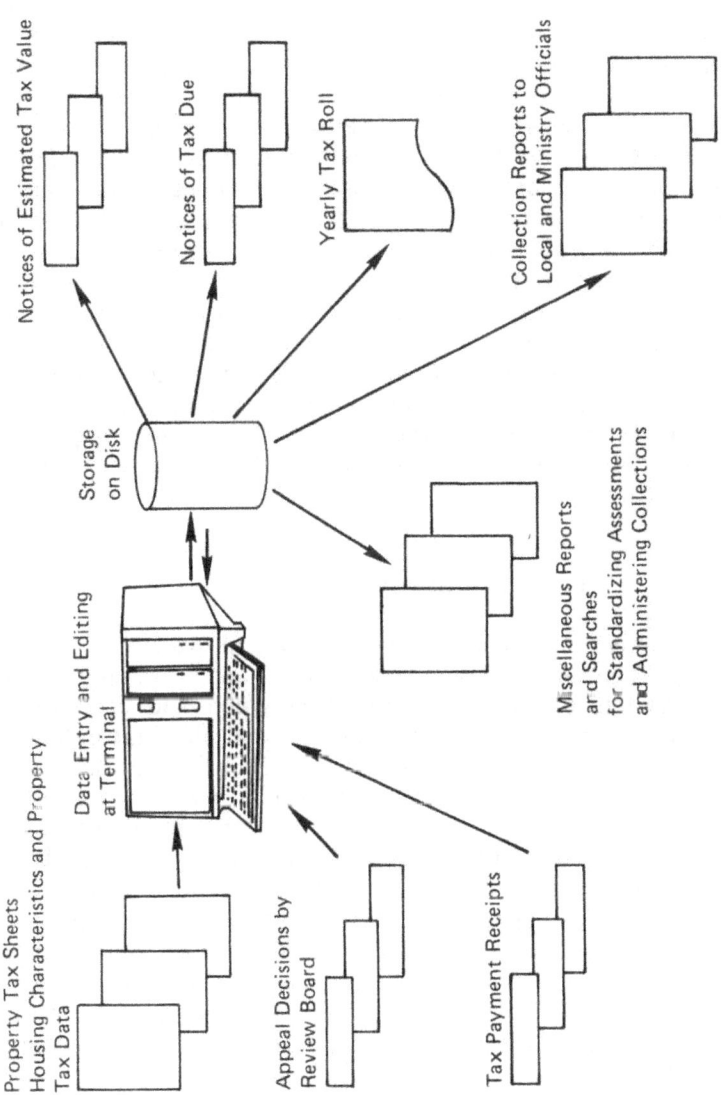

Figure 6-1 Property tax records system

Flexibility was incorporated, allowing new types of files to be created or additional types of information to be included in existing files. The system permits a designated operator to record new information on the files, search for information, print tax bills, and produce special reports. Training requirements are minimal, since the system is "menu driven," that is, menus appear on the computer screen that direct the operations that can be carried out. However, for reprogramming the system (for example, creating a new type of file) more specialized programming knowledge is required.

The system is relatively inexpensive and robust. It can handle up to 20,000 records, for a total hardware cost of about $10,000. In addition, there are modest costs of training local staff and customizing the software for the specific application. The time required to implement the system varies. A major time cost is in entering the information on individual properties into the system; this varies according to the amount of information recorded on each property and the number of properties in the system.

In Tunisia, the prototype is being pilot tested in one city to be fully implemented in a second city. The plan is to introduce it into all secondary cities across the county. The impact expected in Tunisia is that property tax collections will double when the systems are operational.

Geographical Information Systems

Assessment and management of natural resources, land use planning, urban and regional development, and other key development activities often require the acquisition and processing of large volumes of alpha-numeric, graphic, and pictorial data, much of it location and time dependent. Collecting and presenting this data in a form that can be readily manipulated and assimilated while preserving its temporal and spatial attributes presents many challenges. Special information technologies are capable of handling these needs, often with the help of microcomputers.

The advent of digitization, the conversion of all types of data (text, numbers, graphs, pictures, images, sound, spectra) into digital "bitstreams," now provides a flexible, universal means of transporting and manipulating data. When coupled with fast, low-cost microprocessors and storage devices and new techniques for image processing and display, it becomes possible to develop multipurpose information systems that can manipulate, integrate, generate, and display large and diverse arrays of geographic, environmental, or other data in totally new ways and at relatively low cost. Real-time simulation and modelling, detection and depiction of interrelationships and inconsistencies in data, improved data management, and a host of other techniques are now feasible.

The current state of the art reflects the integration of research and technical developments in several areas, including:

- Optoelectronics and optomechanics
- Image processing (encoding, enhancement, compression, understanding, restoration and analysis)
- Spatial analysis and display
- Video technologies
- Computer graphics
- Computer-aided design and drafting (CADD)
- Photogrammetry
- Digital cartography and computer-mapping
- Mass memory and storage technologies
- Mathematical, modelling, algorithmic, and programming techniques.

Institutions involved in advancing the state of the art and its technological embodiment are equally diverse and include U.S. government agencies such as NASA and USGS, academic institutions such as Cal Tech's Jet Propulsion Lab and Harvard's Laboratory for Computer Graphics and Spatial Analysis, and private companies ranging from IBM to small applications consulting firms and equipment manufacturers. Anticipated developments, including the manufacture of microprocessor-based special purpose machines and turnkey systems that are smaller, less expensive, more powerful, and user friendly, make it feasible to consider the routine use of these technologies by developing countries within the next decade.

One of the more important areas of application of this technology for municipalities is in geographic information systems (GIS), which allow the integration and manipulation of geographic, socioeconomic, and statistical data, from maps, tables, graphs, photos, and other images, to produce mapping displays suitable either as cartographic end products or for additional analysis. Common characteristics of GIS data are that they are location specific, measurable, and subject to scientific classification. If in each category data is visualized as an overlay of a base map, then each class of input data can be extracted for a specific location. Thematic searches can also be made to identify particular classes in a specific geographic unit (for example, all commercial forest areas within a given country).

This technology is also useful in energy resource applications involving large volumes of physical and geochemical data. Image processing and related techniques, for example, have been used in monitoring forest defoliation, geological and soil cover, water resources, and power line corridors.

Well-considered, well-constructed spatial databases are vitally important in developing viable geographic and energy information systems. A nontrivial and continuing problem is that of assembling the desired data and restructuring it into a standardized, computer-compatible form. The functions of data collection and development thus dominate a large portion of the activities of operational systems. Much of the physical data appropriate for these systems can be obtained by remote sensing technology, which holds great promise in filling the spatial data void in developing countries (National Academy of Sciences, 1977).

A few developing countries are beginning to utilize this technology in municipal management. Indonesia and Malaysia have operational GIS systems, and other candidate countries, identified in a recent World Bank report as having both the necessary data and the administrative and technical infrastructure capable of applying the technology, include Brazil, Chile, Colombia, the Fiji Islands, Korea, Mauritius, Mexico, Paraguay, Peru, the Philippines, Swaziland, and Zimbabwe.

Microcomputer Technology in Municipal Management

It is still the early stages of microcomputer development for municipal management, particularly with respect to secondary cities in Africa. What are the problems of implementing this new technology and by what means should one proceed?

There is clearly a need for institutional support, within both private and public sectors. Hardware and software vendor networks need to be established. Critical service and maintenance must be available as well as replacement parts. While the requirements of microcomputers are not as large as those of larger computer systems, they do exist and must be addressed, particularly for crucial record keeping systems that cannot be allowed to stay down for long periods.

Assistance in developing customized applications and training also needs to be provided. In many countries, this is best done through specialized institutions that already exist to serve local government. For example, in implementing the tax records system in Tunisia, RTI is working both with the Ministry of Interior and with the National Computer Center, which provides software development and training in computer systems to local governments. In another example, RTI is working through the Brazilian Institute of Municipal Administration (IBAM) to develop the software for a microcomputer model to analyze the impact of municipal investments on local government financial status. In this case, IBAM will be responsible for disseminating the model throughout Brazilian municipalities.

A final issue in implementation is financial: how to pay for this new technology? We see a natural strategy developing in which the acquisition of the new systems will help create new revenue for the local governments. Experience shows that improvements in tax records or utility billings can pay for the systems and begin to generate additional revenues. Municipalities should look first at these revenue-generating applications that can fund additional management improvements.

PART III

Major Policy Issues
and the Future

7

Microcomputer Policy Development: A Case Study of Sri Lanka

The transfer of microcomputer technology to developing countries has scarcely begun, but the process is gaining momentum. Uncontrolled dissemination of technology can be as disruptive as it is beneficial. It is therefore important in the early stages of the process to step back and determine what problems may arise and how they can best be avoided.

This chapter introduces issues that should be addressed by decision makers involved in the transfer of microcomputer technology to developing countries. Many of these ideas were discussed at the International Symposium on Microcomputer Applications held in November 1984 in Sri Lanka. While applications in health, agriculture and energy were stressed, policy issues of particular interest to Sri Lanka quickly came to the fore. Those policy issues form the background to this chapter.

THE ROLE OF MICROCOMPUTERS IN DEVELOPMENT

"The information revolution will some day be seen by historians as having been as fundamental in its human impact as the two great convulsions that preceded it, namely the agricultural and industrial revolutions," said Stanford anthropologist Robert Textor in initiating discussions at the symposium. Textor went on to predict that because of very high data storage, low cost, low energy requirements and portability, microcomputers will become the most common pieces of technology in the world, useful to all sectors of the economy. He suggested that ethnocentrism and especially tempocentrism (the assumption that the values and experience of our time generalize to other epochs) greatly complicate forecasting and planning for the information revolution, but that the profound economic, cultural, and psychological impacts to be expected make policy planning a necessity (Textor, 1984).

Before substantive policy issues are considered, the question of how to make microcomputers serve development should be addressed. Development implies that people's quality of life will be enhanced; what constitutes a full and prosperous life varies from culture to culture, raising two serious problems.

- Those who make decisions about introducing microcomputers will often be from cultures or subcultures different from those whose lives are being affected. Later, even if innovation is considered technically "successful," the success might be definable only in the innovator's terms. Or, both sides might agree that the immediate goal of innovation was successful, but the secondary effects, in terms of the values of the receiving populace, might be undesirable.
- The innovation might produce primary and secondary results that are successful in terms of the values of both innovators and receivers, yet benefit only some receivers, leaving others no better off, or even worse off, than before. The case of the "Green Revolution" is relevant. It tended to benefit farmers who were already better off and hence could afford the fertilizer and other inputs. Poor farmers sometimes became relatively poorer.

The following questions should be considered in any technology transfer process:

Can a New Form of North-South Dependency be Prevented?

Granted that the technologically developed world must, at this stage of history, provide much of the microcomputer technology, can we conceive and design ways in which this technology will reduce, rather than encourage, technological and hence other forms of dependency on the economically advanced nations? Microelectronics should be used to enhance autonomy, not domination.

Can Cultural Distortion be Avoided?

Experience to date with imported computer games, to cite but one example, suggests that they can be culturally distorting when used by people in the Third World, especially young people. The government of Singapore, among others, has responded to this problem simply by banning computer game arcades. However, cultural distortion can occur in more subtle ways, such as through imported educational materials. In the case of both dependency and distortion, perhaps the most promising longer-range solution lies in encouraging and supporting Third World nations to develop their own software and courseware. Such an autonomous approach can be promoted as a matter of explicit national and international policy.

Can Equity be Ensured?

No society is perfectly equitable, but development involves redressing gross inequities and moving toward a society in which the luxuries of the few will not be predicated on the denial to the many of the basic necessities of a decent life. The problem of much technology transfer from industrial countries during the past four decades of technical assistance is that it was introduced in such a way that the rich became richer and the poor became poorer.

Development requires that there be incentives to motivate hard work and frugality, and it is recognized that an incentive system implies some degree of inequality of reward. Inequality of reward, however, need not imply inequity of reward. Can we conceive and design ways of introducing microcomputers so as to promote the kind of development that provides incentives for creative innovation, while at the same time fostering economic equity?

Can the Pain be Shared?

The very best of planning and participation will never be adequate to forestall all difficulty and pain. The widespread introduction of microcomputers in developing countries, it is clear, will affect the responsibilities of many workers. However, policies will doubtless vary enormously in the degree of responsibility shown toward those who no longer have the skills required to carry out their jobs. Subsistence support and occupational retraining will certainly help ease the pain, and this should be part of every major plan.

Why Developing Countries Need Computers

Computer technology is changing the world within which developing countries struggle for economic survival and growth. This change is a matter of great significance, because computers can profoundly affect the central social and economic functions of an interdependent world economy. This technology will change every country's development opportunities. How developing countries manage the computer technology change process will influence whether their development goals will be achieved. Structuring that process will determine who will benefit and in what ways, and will require systematic consideration in the formulation and implementation of national policy (Munasinghe and Blankstein, 1984).

A great deal has been said about the declining costs and ever-increasing capabilities of microcomputers. Great strides continue to be made in technology development and wider use by the public in countries like Japan, France, and the United States. However, in developing countries that do not have easy access to this technology, governments must begin to formulate a policy that will encourage the use of microcomputers in their development. Imported computers are frequently much more expensive in developing countries than in the country of origin because of import duty, high dealer mark-up to make profits in a small market, and high service costs.

Few, if any, developing countries can exploit computer technology opportunities without a number of complex decisions. Such decisions are not necessarily easy to arrive at in countries beset with a multitude of economic problems and constraints. For example, it is difficult to think about a "computer sector" of a developing country economy in the same way one envisions an agricultural or energy sector. Economic data is not organized to facilitate the observation of developments in the computer sector.

MAJOR POLICY ISSUES

Before considering the substance of the policy issues, we should define just what constitutes a "policy" issue. For these purposes, a policy issue is one bound up with public concern and deliberate public action. The "right" policy decision does not always lead to public intervention. Indeed, microcomputer technology has developed largely without public policy intervention. The technology has developed so quickly that there has been little time even to investigate the impact on the common welfare, let alone formulate common action (McCullough, 1984).

Many would say that this benign neglect on the part of government has been a main strength of the microcomputer phenomenon and responsible for its rapid technical advances. Therefore, part of the success of this so-called revolution can be traced to an absence of public policy.

Perhaps the easiest way to approach these problems is to begin with specific technical issues and proceed to the broader international issues.

Technical Issues

Standardization

Computer technology, because it is in a state of flux, has until now defied most attempts at standardization. Data storage formats, for instance, may vary widely from one machine to the next, and programs written for one computer will not function on another. The result, unfortunately, is redundant effort and expense: programs must be rewritten for different systems and data cannot always be shared among users. Even the practical knowledge gained in working with one computer or software package may be useless when applied to another, resulting in a need for additional training. Standards in the computer field are typically imposed by the marketplace, as consumers "vote" for the superior system with their purchases. This process is slow, however, and may not produce practical standards for some years or even decades to come. Can such standards, instead, be produced by international agreement? If so, is it desirable to do so? Will artificially imposed standards stifle desirable innovations in computer technology? It is possible to create flexible standards that foster software data portability while still allowing for technological innovation (that is, operating systems and high level programming languages that minimize differences among machines).

Standardization should promote some degree of efficiency in setting up maintenance and support systems. It allows a smaller foreign exchange outlay for the spare parts and software that must be imported. Standardization also means that users within a country will be better able to share experience. On the other hand, standardization may have drawbacks if the wrong standards are imposed. And in such a rapidly changing industry, premature standardization may close off beneficial avenues of development.

Service and Maintenance

The absence of service and maintenance facilities in many developing countries is a serious problem and may, in some instances, prove to be an insuperable barrier to the effective transfer of microcomputer technology. Can manufacturers be encouraged to locate these facilities in countries? Do current government policies in some countries actively discourage manufacturers from establishing these facilities?

The more we rely on microcomputers, particularly for time-sensitive data handling, the more we need support systems to keep them running. Currently, in many countries the basic computer hardware may be available, but spare parts inventory is not. What will be necessary to develop such support networks? What existing government policies foster, or inhibit, such development?

Power Supply and Communication

The state of earlier technologies that exist in developing countries will also have a strong bearing on the ease with which microcomputer technology may be assimilated. For example, uneven electrical power quality can limit the reliability of microcomputers and increase the cost of a microcomputer system (by necessitating additional equipment to ensure a constant power supply and to protect hardware from electrical surges). The quality of local telephone service will determine the feasibility of high-speed long distance data transmissions. Can existing power supplies in developing countries be upgraded in light of the power quality requirements of microcomputers? Can microcomputer systems be made available that are less susceptible to power fluctuations, or that can be powered inexpensively in the absence of reliable local power?

Microcomputers are also finding an important role in communication and data transfer. The once distinct fields of communications and computers are merging to the point where the telephone companies sell personal computers and computer companies sell phone services. Satellite communication is bringing long distance data transfer closer, making it possible to transmit data from one microcomputer to another located on the other side of the planet. However, to make use of this potential requires the critical link-up over local telephone lines. These local systems currently represent the weak link and serve to limit the communications potential of microcomputers in the near future.

Patents, Copyrights, and Licensing

The development of microcomputers has been almost exclusively a private sector venture, responding to market forces. Consequently, the spread of microcomputer technology will depend largely on international market forces and the business climate within individual countries. While individual governments may enter into special arrangements with computer firms, it will be the commercial decision makers of the firm who determine their willingness to participate. At the same time, the growing competitiveness in the worldwide microcomputer market is already sending computer entrepreneurs overseas in search of new opportunities. Will the entrepreneurial nature of microcomputer firms make the transfer of this technology different from other technology transfers?

National Issues

Equipment Manufacture

Should nations promote the development of their own micro-computer industries? This question is frequently asked in many developing countries. Certain industries have historically been considered essential to national economic or security interests. Should national governments promote the microcomputer industry in the same manner, and to what extent should they, or could they, protect the industry? Indeed, how does a government protect an industry at a stage of development where the direction of the technology is not yet firmly set? Would protectionism choke off the influx of new ideas, which seem to be critical to the current development of the technology? Furthermore, there are many individual components of the microcomputer industry, and the production of hardware constitutes only one part. What aspect of the industry would a national government promote, and who would make that decision?

Ownership and Access

Who is allowed to own and use microcomputers? Many countries have foreign exchange shortages and policies regulating the classes of goods that can be imported. In these cases, government regulation already determines who can import microcomputer systems and who will have access to them. Will usage be restricted to government agencies or state companies? Even within government, will government agencies continue to maintain central control over data processing, or will microcomputer technology allow decentralization? If government restricts access to micro-computers, will parallel markets emerge selling hardware and software at inflated prices? This might restrict access to the wealthy and keep local markets from growing.

Labor Displacement

Will microcomputers displace labor in local economies? This issue has both long- and short-run implications. In the short run, the introduction of microcomputer technology is likely to increase job opportunities, as new types of information-processing activities are made possible. On the other hand, over the long term, micro-computer technology, particularly robotics, may have a profound impact on the structure of work. Bound up with this is the issue of the labor wage competitiveness of developing countries in the world economy as the field of robotics develops. Can these impacts be foreseen with enough clarity to allow policymakers to evaluate them? Can anything be learned from the changing labor structures of the more developed countries as they enter the so-called information age?

Centralized Authority and Institutions

What will be the impact of microcomputer technology on centralized authority? As noted earlier, microcomputers can extend computerization to lower levels of government, business, and even to individuals who have not had access to computers before. Since information is power, this represents a potential power shift, at least within the government bureaucratic structure. For national governments wishing to decentralize, microcomputers provide very useful and powerful tools. However, for governments trying to maintain central control, the very nature of the technology may serve to undermine that tight control.

Can existing institutions deal competently with the technical and policy issues raised by the microcomputer revolution? Most countries have organizations mandated to deal with issues of science and technology. Are these organizations capable of dealing with such a rapidly developing field, a phenomenon that does not allow much time for study and reflection? Some governments are tempted to place a moratorium on the acquisition of this technology until they can study it more closely. Is this wise, or even possible?

Sociocultural Impacts

Microcomputers, because they serve in many instances to ease the burden of the workplace, raise the specter of job displacement, an issue of considerable importance in areas where unemployment is already high and jobs are often precious. Do microcomputers, or computers in general, increase unemployment? Do they displace workers? In offices where microcomputers are used as word processors, for example, is the need for clerical help diminished by the reduced need for retyping documents? Conversely, could microcomputers actually create new jobs, by introducing a need for operators, programmers, and service personnel? What are the required skills and training?

Do microcomputers disrupt cultural and social structures within developing countries? Because microcomputers and software are by and large designed and manufactured in developed countries, could they reflect, and reinforce, sociocultural biases that are alien to developing countries? Is it possible to gauge the impact of such biases within those countries? Could the impact of these sociocultural biases be beneficial in some instances, rather than disruptive?

International Issues

Transborder Data Flow

Transborder data flow, the movement of valuable information resources between nations, has become an issue of considerable significance in light of advances in communications technology and the presence of transnational corporations in the developing world. Who owns data? Do developing countries have the right to control information vital to their interests, even if corporations, individuals, or institutions claim a similar interest or a prior right? These questions are made more immediate by the ease with which computers, including microcomputers, can manipulate, store and transmit information. Will the proliferation of computer technology reduce the ability of countries to control the flow of information across their borders?

Who will control the flow of data within a country and across national borders? This issue is closely related to the access issue noted earlier. Microcomputer technology makes the exchange of data very easy and cheap. In the United States, microcomputer users form extensive networks for the exchange of information and software, sometimes violating copyright laws. At the same time, this free flow of information has been central to the development of new software and new applications, greatly extending the usefulness of the technology.

Products in the International Marketplace

Who determines the type of technology made available to developing countries? This issue concerns both the design of the technology and the flow of technology across national borders. The design question centers on who is the target group of the technology design. Currently available microcomputer hardware and software have been designed primarily for U.S. business and individual consumers. The flexibility required to serve these two groups has produced systems that are also useful in developing country applications, but that has not been the primary consideration. Do current design criteria adequately meet the needs of developing country applications? If not, how uniform are those needs and how should they be met?

Control of the flow of technology across national borders is the second aspect of this broader control issue. Many countries already control the importation of foreign technology. Technology-exporting countries may also move to restrict the outflow of microcomputer technology, as the U.S. Government is now doing on the grounds of protecting national security. To what extent are these controls legitimate and effective? To what extent are they harmful to the technical advance of the technology and to the interests of developing countries?

Do international copyright and patent agreements, or their absence, encourage or discourage computer manufacturers and software publishers from distributing and supporting their products in developing countries? Software publishers in particular might be wary of circulating their products in a region where unauthorized copies might be made and distributed, without their having any hope of legal redress. Would stronger agreements encourage not only the distribution of equipment and software in developing countries, but the development of systems and programs targeted more precisely at the needs of these countries?

New Forms of North-South Dependency

In the absence of a local microcomputer industry (see above), does the introduction of microcomputer technology increase or decrease North-South dependency--reliance of Third World countries on the physical and informational resources of the developed world? Does the use of microcomputers within a developing country create a reliance on software and peripherals manufactured in developed nations? Alternatively, could the use of microcomputers in Third World countries decrease their dependence on the North for information and information-processing services?

Role of the Donor Community

What is the appropriate role of the international donor agencies who are currently funding the purchase of many microcomputers for government agencies? At present, much of the importation of microcomputer systems is done with donor agency funding. At the same time, these agencies are also trying to develop internal policies dealing with the acquisition of computer systems. Do these agencies have a useful and legitimate role beyond funding hardware acquisition?

Many of these above questions were raised at the symposium in Sri Lanka. A number of responses were offered, including a framework for computer policy analysis.

A FRAMEWORK FOR COMPUTER POLICY ANALYSIS

The following framework was developed to facilitate computer policy analysis in Sri Lanka. The purpose of the framework is to place application decisions in a larger perspective. Obviously, not every decision on computer applications has policy implications. Nor should a computer policy, when one exists, control every computer decision. But many decisions can have sub-optimal or counterproductive results if a clear-cut policy is absent. For example, the selection of a language for a program or the choice of a brand of machine for an office can easily be viewed as a purely isolated technical issue. But if the application proves successful and is replicated widely so that it ultimately affects many people, then such choices may have wider significance in terms of training requirements, claims on staff time, and even distribution of financial and social benefits (Munasinghe and Blankstein, 1984).

One can address the problem of analyzing and formulating computer policy in terms of the following classes of influences:

- Major determinants of national computer policy including overall national goals and objectives, constraints that may limit computer dissemination, human resources constraints, characteristics of the commercial sector, entrepreneurial and cultural environment, characteristics of the public sector, the pool of organizations, educational infrastructure, financial resources, and political will.

- Sectors and constraints influencing computer policy options such as characteristics of available technology, the existing national computer base, access to foreign technology, cultural impact, physical infrastructure (for example, power and services), and international relationships with manufacturers.

- Significant issues and problems generated by computer driven change such as employment and job displacement, organizational issues, political dichotomies (modern-traditional, urban-rural), education and language barriers, and technical issues.

There are a number of tools for policy development. These include broad based actions such as direct investment by government, institutional development using extension services, encouragement of foreign involvement through joint ventures and other mechanisms, economic incentive through tax credits, and a variety of regulations in the area of standards and investment controls. Clearly, the choice of which mechanisms to apply will depend on national goals.

In terms of computer policy, these goals can be summarized into five dominant attitudes:

- Complete laissez-faire
- Free market driven, government encouragement through training, procurement
- Large organization driven (essentially as consequence of pricing the technology at a level where only big organizations can buy in)
- Government driven, access and applications controlled
- Security oriented, state controlled.

Preliminary Policy Development

In May 1981, the government of Sri Lanka established the 10-member Computer and Information Technology Council (CINTEC). CINTEC was charged with providing guidance to Sri Lankan public and private sector institutions to help them avoid unnecessary duplication or waste scarce resources of the same time they also had to coordinate controls and regulations to encourage initiative in this rapidly developing field.

Sri Lanka had only a few score computers at the end of the 1970s, and no formal computer training programs; today there are an estimated 2,000 to 3,000 small microcomputers, 500 to 600 large microcomputers, 3 university computer science departments, microcomputer laboratories in 8 of 9 universities, and microcomputers in more than 100 secondary schools. This progress has been achieved under general national guidelines encouraging application of information technology through detailed planning by government ministries, and by allowing considerable initiative on the part of the private sector and individual professionals.

In spite of the process, the use of computers in Sri Lanka is still in its infancy, both in terms of the number of systems installed and their level of sophistication. However, the establishment of CINTEC was based on the conviction that given the support and guidance of the government and a commitment of resources that will be very modest in terms of an overall national investment program, the resulting developments in computers and information technology will bring about incremental improvements in all other sectors of the economy.

National Computer Policy Objectives

The following broad national computer policy objectives were identified in the National Computer Policy Committee's report of April 1983 and subsequently approved by the government of Sri Lanka.

- Harness computer technology in all its aspects, for the benefit of the people of Sri Lanka, and to further the socioeconomic development of the nation.
- Promote and guide the development of computer-related resources and their application in order to anticipate and meet the future needs of the national economy.
- Enhance and supplement manpower resources and increase the efficiency and productivity of management and workers at all possible levels.
- Improve the quality of life of the people of Sri Lanka, including the job satisfaction and working conditions of employees.
- Increase the flexibility and dynamism of Sri Lankan society to enable it to successfully meet the challenges of the future that arise from the ever-increasing pace of worldwide scientific and technological advances.

Policy Guidelines

To meet the broad objectives listed above, the following initial national policy guidelines was drafted. They are revised and updated, as appropriate, on a regular basis.

- Acquisition: Potential users should be encouraged to treat the acquisition of a computer or related items or both as any other investment, including clear-cut identification of computer needs and technical, economic, and financial evaluation of the project. Government imposed regulations, rules, or financial disincentives that would restrict or delay purchasing of computers and related items should be minimized wherever possible.
- Utilization and Access: Sharing of computer hardware, software, and data resources should be promoted. Computer installations should be fully utilized by permitting access to users during as many hours of the day as

possible. However, it would be undesirable and impractical for the government to attempt to compel owners of computer facilities to share their resources. Interchange of information regarding computer hardware and software resources available among different users should be promoted.

- Computer Education, Public Sector Applications, Computer Literacy, and Appreciation of the Potential of Computers: The government should take immediate steps to improve computer-related skills and promote their application as widely as possible, especially in the areas of scientific analysis, higher education, industry, business and financial management, and schools. The establishment of standards for computer education should also have high priority. Particular attention should be paid to identifying and encouraging the application of computers in the public sector. Efforts should be made, as soon as possible, to ensure adequate financial incentives and job satisfaction in order to attract and retain the services of computer personnel in Sri Lanka. Computer literacy and appreciation of the potential of computers among the general public should be increased.

- Self-reliance, Export of Computer Services: Efforts should be made to make the country as self-reliant as possible in computer skills, establish a sound indigenous capability to evaluate and acquire foreign computer technology when necessary, and to export computer services (both software and hardware, especially the assembly of products).

- Computer-related Infrastructure and Legal Environment: The government should give high priority to improving infrastructural facilities that are essential for developing computer use in Sri Lanka, including local and overseas telecommunications services and electricity supply. An adequate legal environment should also be created that recognizes the role of computers as well as their impact on society.

- Other Areas Related to Computers: Developments in areas related to computers, such as satellite communications, other telecommunications, and robotics, should be closely monitored and adapted for application in Sri Lanka whenever appropriate, by both the government and other interested groups.

Computer Development Scenario

A desirable and practical scenario for computer development in Sri Lanka can be examined. In the short run (two to three years), Sri Lanka may expect progressive gains in the productive efficiency of private and especially public sector organizations, through the use of microcomputers in those areas where management skills are scarce. This will enable the intellectual community to enhance its contribution to national development. The initiation of a major effort in computer education, encompassing schools, universities, industry and commerce, and the general public, is already underway.

The medium term (5 to 10 years) is likely to lead to the development of Sri Lanka as an Asian Service Center for computerized international banking and trade. Sri Lanka's assets include the attractive economic policies of the government and stable climate for investment, convenient geographic location, highly educated manpower base, and acceptability among all countries in the region. In this time frame, the development of more decentralized domestic institutions (to meet the needs of administration, finance, production, and exchange of goods and services) can also be expected. This will provide an additional impetus for entrepreneurial activities more in keeping with national character and temperament. By this time, carefully nurtured centers of excellence will be making significant contributions.

In the long run, toward the turn of the century, Sri Lanka should aim for a systematic transformation of the economy. Sri Lanka can move rapidly from the agricultural to the services-oriented stage of economic development, while avoiding some of the worst aspects of the intermediate heavy industrial stage such as environmental pollution and urban slums. Concentration on industries that are knowledge-intensive and efficient in the use of scarce resources is a way to achieve these goals.

Organization of Computer Sector and Policy Implementation

The growth and development of several centers of excellence, such as the Arthur Clarke Center, the universities of Colombo, Moratuwa, and Peradeniya, and the National Institute of Business Management discussed earlier, will be supported. CINTEC is also establishing channels of communication with the Computer Society of Sri Lanka and other private special interest groups and companies. Such nongovernmental bodies will have a key role to play in assisting CINTEC, especially in areas such as:

- Establishing and maintaining a code of conduct for computer professionals
- Maintaining the standards of computer education among private organizations
- Providing a regular forum for exchanging ideas and disseminating information in Sri Lanka
- Helping to ensure the integrity and security of data in computer installations and preventing the abuse of privacy.

SUMMARY

In determining computer policy, a fine line must be drawn between what is considered government regulation and a complete laissez-faire posture. The example of Sri Lanka was given to demonstrate how one country is attempting to rationalize this technological revolution for the good of its people.

Checklists of policy considerations have been proposed. These are presented so that all avenues have been explored in considering impact, opportunity, and limitations. Clearly, each country must chart its own course, but the fact that microcomputers are here and all around us will profoundly alter the way we carry out everyday tasks. For this reason, governments and the private sector must think and plan for this eventuality or else be left behind.

8

The Future

Technology rarely stands still, and microelectronics technology is evolving with dizzying speed. Predicting technological trends is a hazardous venture at best. So rapidly does the field change that predictions are likely to be superseded by the time they have appeared in print. A few broad generalizations can be made about four areas of microelectronics technology that are likely to prove of increasing interest to less developed countries over the next few years: data communications, expert systems, intelligent interfaces, and videodiscs.

DATA COMMUNICATIONS

Data communications is not a future technology; it is available now to any computer user with a modem and access to a telephone system of reasonable quality. There is some question, however, about the applicability of current data communications services for developing countries. This technology was discussed in part in Chapter 3, with particular emphasis on satellites.

New achievements in fiber optics technology threaten satellite based communications, especially in the busy North Atlantic corridor. Fiber optics have been developed to the point where old limitations, which made them economically infeasible, have been overcome. These were primarily bandwidth problems, multipoint capability, susceptibility to storm disruption, and signal dispersion over long distances. However, for the foreseeable future, fiber optics will be limited to busy communication corridors with satellites continuing to handle more remote areas.

EXPERT SYSTEMS

Expert systems, like the intelligent interfaces to be discussed later, are products of a computer science discipline known as artificial intelligence. This field attempts to develop computer programs that simulate certain aspects of human intelligence or mimic behavior normally associated with intelligence. Early successes in this field include computer programs that can play chess and other games of strategy, or communicate in subsets of normal language. Although critics of the field maintain that it has yet to produce results of practical importance, supporters claim that artificial intelligence research has the potential to revolutionize the ways in which we interact with computers.

An expert system is a computer program that "understands" an area of human knowledge in much the same way that a human expert would understand that area of knowledge, which is to say that the program can draw inferences from data on that subject that are not immediately apparent in the data itself. For instance, a simple expert system programmed to understand auto mechanics might infer, given the information that a certain automobile is having difficulty starting, that the car is experiencing problems with its ignition or electrical systems; a somewhat more complex expert system would suggest how this problem might be remedied; and a truly advanced expert system might devise a solution to the problem that would otherwise have occurred only to an experienced mechanic.

The very name "expert system" is a source of some controversy both in and out of the computer field, because it implies that a properly programmed computer with access to a well-stocked base of data might be able to function as well in its area of expertise as a human expert. While this is unlikely, at least in the near future, such "knowledge-based systems" (an alternative term) might serve as a valuable adjunct to human expertise, extending the reach and productivity of the human expert while not precisely serving as a replacement. And, while no computer program may yet be capable of original thinking or profound insight, neither are such programs, if properly used, prone to human sloppiness and error.

Early research in expert systems was performed largely at American universities such as Stanford and the Massachusetts Institute of Technology, both major centers of general research into artificial intelligence. The first such programs, endowed with exotic names like DENDRAL, were employed in fairly esoteric areas of scientific research. (DENDRAL was used in spectroscopic analysis.) In recent years, however, more generalized programs have emerged with applications in areas of interest to developing countries.

Perhaps the most promising use for expert systems is in medical diagnosis. For programmers, the difficulty in capturing the elusive thought processes involved in medical decision making lies in the inexact nature of the data with which diagnosticians must work: a set of vaguely described symptoms from the patient and the results of general medical analyses performed without specific knowledge of which type of disease is being analyzed. Despite these limitations, several effective diagnostic programs have been developed, and the diagnostic results produced by these programs are promising and sometimes quite impressive.

Although these programs are not likely to replace trained medical personnel, and should not replace such personnel if they are available, they do possess important advantages over human experts: they are far more exhaustive than human beings in their consideration of possible alternative diagnoses. In this respect, they serve as an interactive medical library. They can also perform elementary computations, such as the calculation of medicinal dosages, faster and more accurately than can the individual.

Further, they can be used in regions where medical help is otherwise scarce, supplementing human experts in making diagnoses. Thus, they can be of considerable value to developing countries, where medical expertise may be in perilously short supply. However, there is always the danger that the availability of such a system might foster an unwise reliance that could work against a development of more conventional medical diagnostic facilities, and the danger that untrained users could feed in inaccurate information and fail to recognize significantly inaccurate output.

As of this writing, most such diagnostic systems remain experimental, and few are in regular use in clinical situations. As yet, data storage and fast computational speeds required for the effective use of such systems limit the feasibility of diagnostic programs based on microcomputers. However, as microcomputers increase in power, and research in expert systems bears further fruit, medical diagnostic programs of increasing complexity will certainly become available to users in developing countries. Already, the Centre Mondial of Paris has begun field testing, in Chad, a scaled-down diagnostic system based on a British Husky microcomputer.

The usefulness of expert systems extends far beyond medical diagnosis; in fact, any area of human knowledge is fair game for knowledge-based software. The program PROSPECTOR, for instance, is intended to help geologists assess certain types of mineral deposits. Further, generic expert systems are being devised that can utilize properly organized databases containing information from any number of fields and reach conclusions from that data.

An effort is currently underway to develop programs that can write computer programs, thus eliminating the need for human programmers in the development of specialized software. Although so-called program generators have existed for years, they are generally focused within a fairly narrow domain, usually the creation of database programs or report-writing programs (i.e., programs that print organized tabular reports based on the contents of a database). Current research, however, especially within the so-called Fifth Generation computer project underway in Japan, involves the development of more flexible and generalized program-writing programs that will take an unstructured description of a program as input and produce the program itself as output.

Clearly, expert systems represent one avenue by which developing countries could free themselves from dependence on outside expertise, when such expertise is in short supply within the country itself. Although the level of reliance on knowledge-based systems is uncertain at this time, the systems could represent an important supplement to more conventional information resources.

INTELLIGENT INTERFACES

Computers are not as easy to operate as they could be, and certainly not as easy to operate as they should be. A naive, untrained user can rarely interact successfully with one at the first attempt. Most nontrivial software packages, including the operating systems with which the user must interact each time the computer is turned on, require that the operator learn a command language of some complexity, or at least memorize a sequence of keystrokes that can be used to trigger certain desired events. This means, obviously enough, that users must be trained in the use of each software package with which they will be working, or at least must be given time to acquaint themselves with the appropriate materials. This, in turn, adds considerably to the expense and difficulty of implementing a microcomputer system.

This need not be the case. One of the primary differences between computers and other, less flexible, tools is that the essential nature of the computer can be altered through software. A computer is like a piece of clay; if we do not like its current shape, we mold it anew. If it is difficult to use, then we remold it until it becomes easy. The computer need not dictate the terms by which we interact with it; we may dictate our own terms to the computer.

This is a philosophy that computer system designers and software authors are at last beginning to heed. The notion that computers can be made more uncomplicated and easier to use was first explored in detail at the Xerox Palo Alto Research Center (Xerox PARC) in the mid-1970s. These concepts have powered the microcomputer field.

As innovative software (and hardware) designs are increasingly employed to make computers easier to understand, the need for training will decrease, and the expense of implementing a computer system will decrease with it. This will be good news for everyone who uses microcomputers, not just for those in developing countries.

"User interface" is a term that encompasses all of the ways in which a computer and a computer user communicate with each other. It includes the input and output hardware of the computer, the video display and the keyboard, and the disk drives and the printer, as well as the manner in which the software relays its data to the user and accepts data from the user in turn. To reduce the need for user training, it is necessary to make the user interface as transparent and intuitive as possible, that is, to remove the mechanical, difficult-to-use aspects of the interface and replace them with more natural forms of interaction.

An example of such a user interface, incorporating many of the ideas developed at Xerox PARC, can be found on the Apple Macintosh computer, although this type of interface has already begun to appear on other microcomputers as well. The Macintosh deemphasizes the computer keyboard as a method of specifying commands to the computer and emphasizes instead the use of the "mouse," a small handheld box that can be used to position a pointer on the video display of the computer. By pointing at "icons" (computer displayed images representing specific actions that can be performed by the operating system, such as saving files on disk executing programs) and selecting options from "menus" that appear on the screen, the user can, in many instances, make his desires known to the computer without ever touching the keyboard. Because the mouse interface is largely intuitive, it requires little if any training to wield effectively.

The need for training computer users may never be completely eliminated, but further advances in interface technology should go a long way in that direction. The ideal interface would interchange information with the user in the same manner that the user interchanges information with other people—in the user's own language. Artificial intelligence researchers have for several decades sought a means by which computers can communicate in natural languages, that is, human languages as they are used by humans, but have thus far achieved only limited success.

The chief obstacle to implementing true natural language interfaces lies in the ambiguous nature of linguistic structures: words have different meanings in different contexts. Further, human beings often infer the meaning of a sentence from contextual clues that would be unintelligible without previous knowledge of the subject under discussion; computers, having little previous knowledge from which to make such inferences, experience difficulty in following even the simplest of human conversations.

Thus, for a computer to converse with a human being intelligently, it must first have a large base of knowledge and a method of accessing that knowledge very quickly. Although the technical problems involved in implementing a general knowledge base that will allow a computer to converse on equal terms with a human are still somewhat beyond the state of the software art, some work has been done in creating intelligent databases that can converse in subsets of natural language on subjects relating directly to the data that they are maintaining. The user queries the database in simple but natural sentences and the database responds in kind. Over the next decade or two, it is possible that considerable advances will be made in this area; it is a key target of the Japanese Fifth Generation computer project. If so, the need for learning sophisticated command languages may be eliminated altogether.

Keyboarding skills could also be made unnecessary by advances in speech recognition technology, that is, the development of computer input devices that can distinguish spoken words and translate these words into a form that can be processed by a computer CPU. Combined with natural language processing software, speech recognition devices would constitute a powerful and effective user interface. Although this technology is still fairly new, considerable strides have been made in the related field of speech synthesis, the development of output devices that allow the computer to speak with a human-like "voice." Hardware speech synthesizers are already available that can duplicate human speech with uncanny precision; these interfaces are of particular value to blind and illiterate users.

VIDEO DISC SYSTEMS

The video disc represents a new technology for the storage of large amounts of data, visual and textual. For example, a phonograph size disk, about 12 inches in diameter, can store roughly 500 major textbooks or 50,000 slides. This massive read-only-memory provides a substantial database when coupled with an interactive microcomputer system.

The current interest in these hybrid systems, the microcomputer combined with mass storage, is ideal for educational uses. Several firms and institutions have developed entire educational packages for these systems. The U.S. military forces have used this technology for training on a large scale. Their use in developing countries is still in an infant stage. The cost may still be high and the technology not well known. However, this is an ideal vehicle for technical training, particularly in industrial organizations or infrastructure maintenance programs.

Other new technologies that may have a significant impact in the developing countries include robotics and CAD/CAM. These and the other topics previously discussed will be the subject matter of another publication in this series.

Appendix:
Microcomputer Hardware
and Software

The Hardware

A properly functioning general purpose computer system consists of five parts:

1. A method of putting information into the computer (generally termed input)
2. A place to store that information once it is inside the computer (memory)
3. A mechanism for processing the stored information (the central processing unit, or CPU)
4. A means of removing the information in processed form (output)
5. A set of instructions, in a form that can be executed by the computer, describing the specific steps required for the information processing task at hand (the program or code).

The first four parts are referred to collectively as the computer hardware; the program or code is commonly known as the software. The CPU and memory are generally incorporated into a computer system when it is purchased.

The most common input-output device for current microcomputer systems is the keyboard. Modelled after the keyboard on a typewriter, it is commonly used for the input of text and numbers. Other input devices include the disk drive (also an output and memory storage device) and the modem (used to transmit information directly from one computer to another, often via the telephone system). Although the technology is still relatively primitive, there are also devices such as optical digitizers that can process graphic information from a picture, and speech recognition devices that can process, albeit crudely, the spoken word.

The most common display device for microcomputers is the video or CRT, which displays text and graphics on a television-like screen. If a permanent copy of the information is required, rather than the transient electronic image on the video monitor, a printer can be used to produce a written copy (hard copy). Plotters, which are sometimes incorporated into an existing printer, can be used to create hard copies of graphic images, such as bar charts and computer-generated illustrations. The disk drive and the modem can also be used for output.

There are two kinds of computer memory storage—internal and external. Internal storage is directly accessible to the central processing unit; information can be exchanged directly, and almost instantaneously, between CPU and internal memory. Data required for immediate processing by the CPU is stored here, as are the computer programs themselves. Internal storage consists largely of a sequence of circuits in which information can be stored as a series of electronic pulses. One disadvantage of internal memory is that its contents (with some exceptions) are lost if there is an interruption in the power supply—that is, when the machine is turned off. External storage, on the other hand, is only indirectly available to the CPU, but can be used for relatively long-term storage of information; a constant power supply is not necessary. The disk drive is the most popular means of external storage on a microcomputer, though other devices, such as cassette tape recorders and paper punchers, have been used with varying degrees of success or convenience.

Internal memory also comes in two forms: Random Access Memory, or RAM; Read-Only-Memory, or ROM. The first, which could be more accurately referred to as Read-Write Memory, can be used for the temporary storage of information that the user wishes to process; when the data currently stored in RAM is no longer needed, it can be erased and RAM used to store fresh data. ROM, on the other hand, is given its contents at the factory, and cannot be further changed. ROM is therefore used for the storage of programming or data that will be required whenever the computer is turned on and is never altered.

The Software

The final requirement is the program or code. Even the most elementary tasks a computer can perform must be mediated by software, so basic programming is included in ROM.

The central processing unit (CPU) of the computer is capable of performing a large number of basic information processing tasks. The way these tasks are combined is the computer program or code. The program specifies a sequence of instructions and the order in which they are to be executed. A computer program, as executed by the CPU, takes the form of a series of electronic signals. The CPU fetches and executes these instructions from memory in a sequential order.

Programs are created in symbolic notation, which is then translated into electronic form by the computer itself. Myriad symbolic notations are available. The specific language used by a programmer is generally a matter of taste, though certain languages are suited to specific families of tasks. The computer language COBOL (Common Business Oriented Language), for instance, is intended for the creation of programs that process large files of data, whereas Fortran (Formula Translator) is used in scientific work.

Operating Systems

Every computer system must include a software package called an operating system, such as DOS (for disk operating system). As this term implies, this program is usually delivered on a disk, to be read into ROM, though in a few computers it is already available in ROM.

The operating system is the computer's master control program; it mediates nearly everything that the computer does, including the way in which it interacts with peripheral devices, with software, and with the user. The operating system is the first program to execute when the computer is turned on, and often the last to execute before it is turned off.

Each computer uses a particular operating system, though alternative operating systems are sometimes available. Operating systems fall into two camps: proprietary and standard. A proprietary operating system runs on only a single brand of computer; a standard operating system will run on several different brands. In general, the advantage is with the standard operating system, because it allows software portability between computers, thus expanding the base of available programs.

Only a handful of standard operating systems have made inroads into the microcomputer market. The following are the most significant:

CP/M. Developed in the mid-1970s by Digital Research, Inc., CP/M was the first popular standard operating system for microcomputers, and it retains a great deal of that popularity to this day. A wide range of business-oriented software exists, including the popular WordStar word processing package from MicroPro International. One estimate places the number of available CP/M programs at more than 15,000.

MS/DOS. Similar to CP/M in some ways, this operating system from the Microsoft Corporation has supplanted CP/M in the 16-bit world. One version of MS/DOS, called PC-DOS, runs exclusively on the IBM Personal Computer. However, programs intended to run under PC-DOS will also run on other computers using this operating system with minimum debugging.

UNIX. UNIX is a product of Bell Labs, the research and development arm of the American Telephone and Telegraph Company. Originally an operating system for larger computers, UNIX has the advantage of running on more than one computer CPU, thus providing a new order of software portability. Because UNIX was designed for larger computers, it will accomodate unlimited amounts of internal memory. As powerful microcomputers become more widely available, the popularity of UNIX will probably grow and the software base expand.

The operating system is the single most important program run on a computer, but it is purely a means to an end. Unless the intention is to develop all software locally, with the aid of experienced programmers, applications programs will be needed. The range of information-processing applications for which a computer is suitable is nearly infinite, but in practice there are a handful of applications that make up the bulk of all commercial software. The following are a few of the more popular applications:

Word Processing. With word processing software, documents can be prepared and edited directly on the video display of the computer and printed on paper when a physical copy of the document is desired.

In developed countries, computers equipped for word processing have, to some extent, superseded the typewriters for business purposes, especially in the preparation of large documents. The advantage of word processing over more conventional methods of document preparation is that changes in the document do not necessitate a complete retyping; once typing (or "keyboarding") has been performed, only the changes need to be made. New copies may be generated on a printer as often as desired. A typical word processing program allows the insertion, deletion, and alteration of text in a document, as well as sophisticated formatting of the printed copy.

Database Management. A computer, given a high-speed storage mechanism such as a floppy or hard disk drive, can serve as an efficient and intelligent file cabinet for large amounts of information, the "database" for the particular field or area of endeavor. Database management software helps the creation and maintenance of large files of electronically encoded information, generally stored on disk. A typical database package will allow the design and creation of information files, and the retrieval and sorting of the individual records in those files according to certain "keys." For instance, the user may design a file that will contain records of the name, addresses, staff position, and pay levels of employees, and then by inputting data, create a series of files based on this design. Once created, the files can be sorted into alphabetic order according to name or staff position, or into numeric order by pay level. Individual records may be retrieved and printed out according to the information contained in those records.

Electronic Spreadsheets. Unlike word processing and database management software, that have their antecedents in precomputer information processing methods, the spreadsheet is a uniquely electronic way of looking at data (although, as the name implies, it was loosely inspired by the accountant's spreadsheet). A spreadsheet program organizes data as a matrix of cells on the computer display.

The user of the spreadsheet can place data in selected cells, and specify the ways in which the information is manipulated. Then, when the data in one cell is changed in some manner, the information in all dependent cells is updated automatically. Spreadsheets are an excellent method of projecting information such as budgets, because various combinations of numbers can be tested and the results of changing a single number can be seen immediately, allowing complex situations to be modelled and possibilities tested.

Other types of programs include graphics design programs (for the creation of charts, blueprints, illustrations), games, educational packages, and statistical analysis software.

In recent years, a trend has developed toward integrated software packages: programs, or batteries of related programs, that include several applications under a single umbrella, "integrated" together in such a way that the separate applications can share data (that is, numbers from a database can be placed automatically into the cells of a spreadsheet, or vice versa). Commonly, the applications in an integrated package will share a common command structure, so that the user who learns to use one program in the package will have little trouble learning the others.

These programs are often memory hungry and will work on only large microcomputer systems. Lotus 1-2-3, for instance requires a minimum of 256K and the more recent Lotus Symphony will function at its peak only in a system with at least 360K.

In some instances, the cost of software for a computer system, even if only a few major commercial products are purchased, can exceed that of the computer. Many packages are now available in the public domain, that is, without charge. This software is usually created either at universities or by hobbyists and addresses specific needs. A problem sometimes encountered with public domain software is that the written documentation and instructions supplied are inadequate. To offset this, authors are often willing to discuss the problems users encounter. Others provide the software without charge, but ask a small fee for documentation after the user has tested the program.

Because most software is produced in industrial countries, a major problem for developing countries may be the lack of programs that address their specific needs. Thus, serious consideration should be given to local development of modest applications programs. Although large programs represent a substantial investment in terms of money and programmer time, the local production of small programs tailored to specific needs is feasible and may be essential. To this end, however, certain software tools are necessary. As explained, programs are generally written by programmers in high-level programming language. They must then be translated into a form executable by the computer, and appropriate software must be available to perform this translation. Most microcomputers are supplied with the necessary tools for creating programs in BASIC, which is an effective language for the creation of small programs. Further, a wide range of public domain software is available in BASIC, and inexpensive BASIC programs can often be obtained in books and magazines. Software for writing more complex programs in COBOL or FORTRAN, for example, can be obtained if the programmer prefers.

Glossary

BACK-UP SYSTEM
Secondary system, as close to identical to the primary system as possible, to be used when the primary system is not available.

BYTE
The number of adjacent binary digits, usually 8, used to represent one character.

CENTRAL PROCESSING UNIT (CPU)
Unit of a computer that includes the circuits controlling the interpretation and execution of instructions--the "brain."

COMMUNICATION
Transfer of information from one point to another without alteration of its content.

COMMUNICATIONS SOFTWARE (terminal software)
Used in conjunction with a modem, allows the transmission of data, words, numbers, and programs between computers.

COMPILER
Program development tool to prepare a machine language program from a computer program written in a high-level language so that it can be revised for use with available operating systems.

COMPUTER
Data processor that can perform substantial computation, including numerous arithmetic or logic operations, without intervention by a human operator during the process.

CP/M
Operating system used on a large number of microcomputers.

DATA ANALYSIS
Process of comparing data items to generate conclusions.

DATA CLEANING
Program run on data input before it is analyzed to remove items that fall outside a logically defined range.

DATA PROCESSING
Execution of a systematic sequence of operations on a set of data (data analysis, information processing).

DATABASE MANAGEMENT
Type of software program that accepts input information and stores that information on magnetic media (usually either a diskette or hard disk). Once the data is stored, the database management software can sort it into a desired order—alphabetic, numeric, chronological—and retrieve it either in part or in its entirety.

DATABASE
Collection of data fundamental to an enterprise.

ELECTRONIC MAIL
Uses telephone lines to transmit messages directly between computers or between the user and a service that will forward a hard copy of the computer-transmitted information to the recipient.

ELECTRONIC SPREADSHEET
Software program to simulate an accounting spreadsheet. When a change is made in one category, the spreadsheet software will update all dependent factors and immediately display the result. The best combination of factors, and the most effective budget, can be determined.

ERROR RECOVERY
Process of deducing correct input information from exceptions detected by the exception processing program.

EXCEPTION PROCESSING
System designed to detect operator input errors because the items do not fit within certain predetermined parameters.

EXPERT SYSTEM
Form of modelling or simulation designed to incorporate enough information to allow it to mimic human decision making.

FIBER-OPTIC COMMUNICATION
Data communication using light rather than wire as a carrier for information.

FLOWCHART
Graphic representation of tasks to be performed in the completion of a project organized by the time allotted to each task. Sections of these charts can be expanded to show the staff, equipment, and resources needed to perform each individual task.

GENERIC SOFTWARE, see SOFTWARE, GENERIC.

HARD COPY
Copy printed on paper.

HARDWARE
Physical equipment, as opposed to the program (SOFTWARE) or method of use.

IMPLEMENTATION
Choice, purchase, and installation of a system.

INFORMATION PROCESSING, see DATA PROCESSING.

INFORMATION
The meaning that a human assigns to data by means of the known conventions used in their representation.

K
Symbol used for 1024 bytes, a measure of data storage and manipulation capacity.

LINKER
Program development tool that allows programs written in the popular high-level languages to be used with available operating systems.

LOTUS 1-2-3
A commonly used commercial spreadsheet software.

MAINFRAME COMPUTER
Largest type of computer; used for running programs containing large amounts of data.

MASTER CONTROL PROGRAM, see OPERATING SYSTEM.

MEMORY
Name given to the storage area of a computer.

MICROCOMPUTER
Smallest size of computer, ranging from a desktop to a portable machine.

MICROPROCESSOR
An integrated circuit that combines all of the information processing machinery of a computer on a single chip of silicon.

MINICOMPUTER
A midsize computer.

MODEM
MODulator-DEModulator, a peripheral device that is used to transmit electronic data, usually via telephone lines.

MODULAR FORMAT
Style of writing software that allows the modification or replacement of individual parts of a program without the need to rewrite software outside the particular part being replaced.

MS DOS
A computer operating system. A version of this system is used on the IBM PC.

MSTAT
A microcomputer package to design, manage, and analyze agricultural experiments designed at Michigan State University.

NETWORK
A complex consisting of two or more interconnected computing units. They can be permanently connected, or joined over communication lines for a limited time to exchange information.

ONLINE
Connected to a computer or system. Online may be used to describe a user's ability to interact with a system or equipment that is attached to a system.

OPERATING SYSTEM
The master control program usually written in "machine" language and unique for each type of microprocessor.

PACKET SWITCHING
Digital processing technique that allows many digital "conversations" to share the same channel while providing very high-quality communication, thanks to built-in error-checking routines. Each "packet" is a burst of digital data addressed to a specific computer. If not acknowledged, the packet is discarded and the transmitting computer is instructed to repeat the transmission.

PERIPHERAL DEVICE
Any device, such as a printer or modem, that is attached to the computer by means of a cable.

PROGRAM
Series of instructions written to achieve a desired result on the computer.

PROGRAMMABLE CALCULATOR
Simple form of computer for carrying out mathematical operations. The ability to be programmed permits it to be used for repetitive calculations without forcing the user to enter each operation separately each time.

PROGRAMMER
A person trained in the design, writing, and testing of computer programs.

RAM
Random Access Memory, the storage area for data and information in a computer. The information in RAM can be manipulated and changes returned to RAM for storage.

RAW DATA
Name given to facts collected before they have been processed into a meaningful form or information.

REMOTE SENSING
Studying topographical, agricultural, or other information from satellites.

ROBOTICS
The art and science of using machines that mimic human actions to perform tasks.

ROM
Read Only Memory, the portion of computer memory devoted to programming. The items in this memory cannot be manipulated by the user.

SATELLITE COMMUNICATION
Communication carried by geostationary or low-orbit (LEO) satellites. Earth stations are able to send messages, using microwave technology, to the satellite, which then passes the message to the addressee's earth station.

SIMULATION
Representation of certain features of the behavior or functioning of a device or system on a computer.

SOFTWARE COMPATIBILITY
The ability to use a particular piece of software on a particular piece of hardware.

SOFTWARE
A set of programs, procedures, and associated documentation concerned with the operation of a data processing system (contrasts with HARDWARE).

SOFTWARE, GENERIC
Commonly used packages such as word processing, spreadsheet, and database management systems.

SYSTEMS ANALYSIS
Analysis of an activity to determine precisely what must be accomplished and how to accomplish it.

TELECONFERENCE
Meeting conducted by means of a computer. It is also known as an asynchronous conference because the input items stay on the system so they can be received at another time. This contrasts with a telephone conference, called synchronous because all participants must be online at the same time.

TERMINAL SOFTWARE, see COMMUNICATIONS SOFTWARE.

TURNAROUND TIME
Period between the submission of collected data for processing and the return of results to the originating organization.

UNINTERRUPTIBLE POWER SUPPLY (UPS)
A piece of hardware consisting of rechargeable batteries and devices to prevent spikes or surges in the electrical power supplied to the user.

UNIX
A computer operating system designed by AT&T.

USER-FRIENDLY SYSTEM
Hardware/software package designed for the nonspecialist that has incorporated menus, help screens, and other features to make it easy to use.

VALUE-ADDED NETWORK
Central switchboard that automatically connects the desired computer on command. Value-added networks use "packet switching" to provide inexpensive, high-quality communication. Value-added networks may be thought of as pipelines into which many separate data transmissions are combined at one end and emerge at the other end correctly differentiated.

VIDEO DISC
Storage medium that is read by a laser. Its advantages are its large storage capacity and the fact that it can be mass reproduced. The major disadvantage is that it cannot be erased so that new data can be stored on it.

References

Ahmed, S.F. and P.R. Rittermann. Survey of Microcomputer Based Design Tools for Building Energy Calculations, American Solar Energy Society, Anaheim, California, 1984.

Alcala, Jr., L.V. "Interfacing the Data Aquisition Microcomputer to CP/M and MS-DOS Operating Systems," paper presented at the Symposium on Microcomputer Applications in Developing Countries, Colombo, Sri Lanka, 1984.

Aluwihare, A.P.R. "What Can a Doctor do with a Personal Computer?" paper presented at the Symposium on Microcomputer Applications in Developing Countries, Colombo, Sri Lanka, 1984.

American Society of Heating, Refrigerating and Air Conditioning Engineers, ASHRAE Handbook of Fundamentals, ASHRAE, New York, NY, 1970.

Barr, A. and E.A. Feigenbaum. Chapter VIII, "Applications-oriented AI Research: Medicine," in The Handbook of Artificial Intelligence, Vol. 2, William Kaufmann, Inc., Los Altos, California, 1982.

Bennett, J.M., and R.E. Kalman, Eds. Computers in Developing Nations, North-Holland Publishing Company, New York, 1980.

Bertoli, F., and S.C. Bertoli. "Improving Epidemiological Reporting in Morocco: A Report on Use of the Interactive Statistical Inquiry System," American Public Health Association, Washington, D.C., 1981.

Bhalla, A., D. James, and Y. Stevens. Blending of New and Traditional Technologies, International Labor Organization, Geneva, Switzerland, 1984.

Boonthai, M.R. Chalermsook. "Micro-computer for Hospital Management," paper presented at the Symposium on Microcomputer Applications in Developing Countries, Colombo, Sri Lanka, 1984.

205

Broadus, C. R. Hydropower Computerized Reconnaissance Package, Version 2.0, Idaho National Engineering Lab., USDOE, April, 1981.

Cappi, C. and J. Giffen. "DASI, Computer Programmer for Project Analysis," User's Guide for IBM VM/CMS System and Apple II, Food and Agricultural Organization of the United Nations, Rome, Italy, 1982.

Carroll, T.O. Energy Sector Development, Economic Growth and Balance of Payments in Columbia, Report to Estudio Nacional de Energia/DNP, Bogota, Colombia, 1983.

Chao, D.N.W., and K.B. Allen. "A Cost-Benefit Analysis of Thailand's Family Planning Program," in International Family Planning Perspectives, Vol. 10, No. 3 (Adapted from a report of a study conducted as part of the Integrated Population and Development Planning Project by RTI funded by Office of Population), Agency for International Development, Washington, D.C., 1984.

Christie, A. "The Problem of Wheat." SOFTALK, 1984.

Cole, H.E. and D.J. Edelman. "The Futures Group Fuelwood Model," The Futures Group, Washington, D.C., 1983.

Daly, J.A. Appendix II, "Comparing Risk Formulas: A Structural Approach," Mathematical Techniques in Health Administration, Ph.D. Dissertation, University of California at Irvine, 1985.

Daly, J.A. "Microcomputers, Risk Indicators and Health," paper presented at the Symposium on Microcomputer Applications in Developing Countries, Colombo, Sri Lanka, 1984.

Davis, G.B. Management Information Systems: Conceptual Foundations, Structures, and Development, New York: McGraw Hill Book Company, 1974.

Development Project Managment Center. "Microcomputers and Agricultural Management in Developing Countries," Proceedings from the Practitioner Workshop. USDA/OICD, Office of International Cooperation and Development, Washington, D.C. 1982.

Diskin, B., et al. "Considerations for Use of Microcomputers in Developing Country Statistical Offices," International Statistical Programs Center, U.S. Bureau of the Census, Department of Commerce, Washington, D.C., 1983.

Eddy, A., et al. "Documentation of Software Developed for the Use of Micro Processors in Managing Climate and Weather Data," Oklahoma Climatological Survey, 1984.

El Kholy, A., and S.H. Mandil "Microcomputers and Health Improvement in Developing Countries," in WHO Chronicle, Vol. 37, No. 5, 1983.

Etherington, D. and P. Matthews MULBUD User's Manual. Development Studies Centre, The Austrialian National University, in collaboration with International Council for Research in Agroforestry, Canberra, Australia, 1984.

Evans, C. The Micro Millennium, New York: Pocket Books, 1979.

Feldstein, M. "A Binary Variable Multiple Regression Method of Analysing Factors Affecting Peri-Natal Mortality and Other Outcomes of Pregnancy," Journal of the Royal Statistical Society, Series A., pp. 61-73, 1966.

Fisher, R.A. "The Use of Multiple Measurements in Taxonomic Problems," Annals of Eugenetics, Vol. 7, pp. 179-188, 1936.

Foell, W., et. al. "Preliminary Report of the Intensive Course/ Workshop on Energy System Analysis and Planning in Indonesia," Energy Research Center, University of Wisconsin, Madison, 1983.

Freed, R.D., and M.T. Weber. "A Progress Report on MSTAT--An Agricultural Research, Design, Managements, and Analysis Software Program for Microcomputer Use in Developing Countries," paper presented at the Symposium on Micro-computer Applications in Developing Countries, Colombo, Sri Lanka, 1984.

French, D.M. "Origin and Development of The Strengthening Health Delivery Systems (SHDS) Approach to Data Management in the Field," Conference on Computer Technology and International Health, Washington, DC, January 25-26, 1984.

Frerichs, R.R. and R. Miller. "Introduction of a Microcomputer for Health Research in a Developing Country--the Bangladesh Experience," Public Health Reports, Department of Health and Human Services, Washington, D.C., November/December 1985.

Fritz, J. Small and Mini-Hydropower Systems, New York:McGraw-Hill, 1984.

Gadgill, A. "Microcomputer Model of Thermal Performance and Energy Consumption of a Residential Building," paper presented at the Symposium on Microcomputer Applications in Developing Countries, Colombo, Sri Lanka, 1984.

Gadgill, A., G. Gibson, and A.H. Rosenfeld. TWOZONE Users' Manual, CBC Report 6840, Lawrence Berkeley Lab., Berkeley, California, 1978.

Gillings, J. "Designing and Implementing National Energy Conservation Programs," Acres International/CDA Case Study, Dakar, Senegal, 1984.

Gordon, M. "EnVest: A System for Planning and Investment Analysis," Development Science Inc., Sagamore, Massachusetts, 1983.

Gordon, M., et. al. "Energy Planning in Morocco," DSI Technical Report, Development Sciences Inc., Sagamore, Massachusetts, 1984.

Gulliver, J.S. "Hydropower Surveys with Microcomputers," paper presented at the Symposium on Microcomputer Applications in Developing Countries, Colombo, Sri Lanka, 1984.

Gunawardena, J.A., K.B.N. Ratnayake and D.S. Wijesundera. "Linkage of Clinic Records Containing Inconsistent Data," paper presented at the Symposium on Microcomputer Applications in Developing Countries, Colombo, Sri Lanka, 1984.

Hebert, P.V. "Computerized Techniques for Improved Planning and Design of Water Supply Systems," paper presented at the Symposium on Microcomputer Applications in Developing Countries, Colombo, Sri Lanka, 1984.

Helfenbein, S. "Training for Use of the Micro-Computer for Data Management in the Field," Unpublished paper, project for strengthening health delivery systems in Central and West Africa, Boston University, Boston, Mass., 1984.

Hsieh, L. "A Study on the Microcomputer Programming for Feed Formulation," paper presented at the Symposium on Microcomputer Applications in Developing Countries, Colombo, Sri Lanka, 1984.

Ingle, M.D., and K. Smith. "Microcomputers and Agricultural Organizations: Management Applications in Developing Countries," Inter-American Institute for Cooperation on Agriculture (IICA) Seminar on Microcomputers and Agricultural Development in Latin America, San Jose, Costa Rica, 1983.

Lauffer, S. and M. Anderson. "A Study of Access to On-Line Databases from Latin America and the Carribbean," Academy for Educational Development for the U.S. Agency for International Development, Washington, D.C., March, 1984.

Lawless, W. and S. Passman. "Informatics and Tunisian Development," U.S. Agency for International Development, Washington, D.C., 1985.

Lokmanhekim, M., et. al. "DOE-2: A New State-Of-The-Art Computer Program for Energy Utilization Analysis of Buildings," Lawrence Berkeley Lab. Report CBC - 8977, Berkeley, California, 1979.

Majumder, A.K. "Computer Analysis of Frequency Spectrum of the Phonopulmogram," Proceedings of the 12th Annual Conference of the Society for Advanced Medical Systems, Vol. 1, Washington, D.C., November 1-5, Published by IEEE, Piscataway, N.J., 1980.

McCullough, J.S. "Policy Issue Definition," paper presented at the Symposium on Microcomputer Applications in Developing Countries, Colombo, Sri Lanka, 1984.

McCullough, J.S. "Microcomputer Technology for Municipal Management," Unpublished paper, 1985.

McGrann, J.M., et al. "Microcomputer Budget Management System," College Station, Texas: Texas A&M University, Department of Agricultural Economics. (Mimeograph). 1984. (This software development effort was also reported upon in 1982 at the MSU Conference on Microcomputers in International Agriculture: See Chapter VII-F, pp. 85-88 in Weber, et al., 1983.)

Milstein, J. "Use Your Computer As A Solar Design Tool," Popular Science, December, 1982.

Munasinghe, M. and G. Schramm. "Energy Economics," Demand Management and Conservation Policy, New York: Van Nostrand Reinhold Co., NY, 1983.

Munasinghe, M. "Integrated National Energy Planning (INEP) in Developing Countries," Natural Resources Forum (4), p. 359; also available as reprint No. 165 from the World Bank, Washington, D.C., 1980a.

Munasinghe, M. "An Integrated framework for Energy Pricing in Developing Countries," The Energy Journal p. 1; also available as reprint No. 148 from the World Bank, Washington, D.C., 1980b.

Munasinghe, M., and C. Blankstein. "Computer Policy Framework and Issues in Developing Countries," paper presented at the Symposium on Microcomputer Applications in Developing Countries, Colombo, Sri Lanka, 1984.

Munasinghe, M. "Welcoming Address," paper presented at the Symposium on Microcomputer Applications in Developing Countries, Colombo, Sri Lanka, 1984.

Munson, J.S. and P.F. Palmedo "The Use of Microcomputers in Energy Planning for Developing Countries," Energy/Development International Report, Setauket, N.Y., April 1983.

Munson, J.S., P.F. Palmedo, S.B. Dhar, R. Nathans, B.G. Tunnah. "The Use of Microcomputers in Energy Planning for Developing Countries," Final Report, Office of Energy, USAID, Washington, D.C., 1983.

Murphy, B. "The Issue is Access—or Lack of It," African Business, June, 1984.

National Academy of Sciences. "Resource Sensing from Space," National Academy Sciences, Washington, D.C., 1977.

National Academy of Sciences. "International Workshop on Energy Survey Methodologies for Developing Countries," National Academy of Sciences, Washington, D.C., Jan. 21-25, 1980.

210

Nelson, E.E., J.A. Daly, and R.D. Joseph. "Time Varying Threshhold Logic," Chapter 12 in Biophysics and Cybernetic Systems, Spartan Books: Washington, D.C., 1965.

Olson, C.D. "High technology is sometimes the appropriate technology for health care in developing countries," unpublished article, Sept. 1984b.

Olson, C.D. "Microcomputers and health: information systems in Lesotho," unpublished article, Sept. 1984c.

Orth, H.M. "Microcomputer Applications in Environmental Engineering at the Asian Institute of Technology," paper presented at the Symposium on Microcomputer Applications in Developing Countries, Colombo, Sri Lanka, 1984.

Perlack, R.D., W. F., Barron and S. Das, "A Microcomputer Model for Evaluating the Economic Prospects of Wood Energy Plantations," paper presented at the Symposium on Microcomputer Applications in Developing Countries, Colombo, Sri Lanka, 1984.

Phillips, J.F., D. Leon, and A.B.M. Alam Mozumder Korshed. "The Design of a Sample Registration System for Monitoring Demographic Dynamics and Health and Family Planning Service Operations in Rural Bangladesh," paper presented at the Symposium on Microcomputer Applications in Developing Countries, Colombo, Sri Lanka, 1984.

Pinckney, T.C., J.M., Cohen and D.K. Leonard. "Microcomputers and Financial Management in Development Ministries: Experience from Kenya," Report to Harvard Institute for International Development, Harvard University, Discussion Paper No. 137, 1982.

Radloff, S. "Microcomputer Applications in USAID'S Population Sector Projects," private communication, 1985.

Ratnayake, C. "Computer Applications for Power System Distribution Study," paper presented at the Symposium on Microcomputer Applications in Developing Countries, Colombo, Sri Lanka, 1984.

Sharma, P. "MSTAT: A Microcomputer Software Program for Agricultural Research," paper presented at the Symposium on Microcomputer Applications in Developing Countries, Colombo, Sri Lanka, 1984.

Shires, D.B. "Health Information Systems Application in Technologically Lesser-Developed Countries: An Overview," report prepared for the Canadian International Development Agency (CIDA), Ottawa, 1982.

Sox, H.C. Jr. "Examining the Medical Decision Making Process," Dartmouth Medical School Alumni Magazine, pp. 20-24, Spring, 1984.

Srivastava, A., and V.P. Sharma. "Simulated Model of Seasonal Effect on Malaria Situation," paper presented at the Symposium on Microcomputer Applications in Developing Countries, Colombo, Sri Lanka, 1984.

Stillwell, T.C., et al. "The Agricultural Statistical Analysis System in Microcomputers and Programmable Calculators for Agricultural Research in Developing Countries," International Development Working Paper No. 5, by M.T. Weber, et al., East Lansing, Michigan: Michigan State University, Department of Agricultural Economics, 1983.

Strain, J.R. and S. Simmons. "The Cooperative Extension Service Updated Inventory of Agricultural Computer Programs," Circulars 531-A, B, and C, Gainesville, Florida: University of Florida, Institute of Food and Agricultural Sciences, Food and Resource Economics Department, 1984.

Strong, M.A. "A project to develop microcomputer software from demographic research: Initial suggestions," Unpublished paper, April 15, 1983.

Teel, J.H. and R.K. Ragade "A System Dynamic Approach to Family Planning and Health Services in Developing Countries: the Bangladesh Case," 1983 World Conference on Systems, Caracas, Venezuela, July 11-15, 1983.

Teoh, S.-T., and R.I. Rasiah. "The Use of a Microcomputer System at District Level Health Services," paper presented at the Symposium on Microcomputer Applications in Developing Countries, Colombo, Sri Lanka, 1984.

Textor, R.B. "Shaping the Microelectronic Revolution to Serve True Development," paper presented at the Symposium on Microcomputer Applications in Developing Countries, Colombo, Sri Lanka, 1984.

Tohar, D. "Use of Microcomputers in Agricultural Research Institutes with special reference to Sukamandi Food Crops Research Institute, West Java, Indonesia," paper presented at the Symposium on Microcomputer Applications in Developing Countries, Colombo, Sri Lanka, 1984.

Viswanath, A. "Optimization of Wastewater Treatment and Sludge Disposal Systems: A Package for Micro-Computer Application," Research Study, Environmental Engineering Division, Asian Institute of Technology, Bangkok, Thailand, 1984.

Weber, M.T., et al. "Microcomputers and Programmable Calculators for Agricultural Research in Developing Countries," MSU International Development Working Papers, No. 5. East Lansing, Michigan State University, Department of Agricultural Economics, 1983.

Whitehorne, E.W., and D.A. Trottier "A Microcomputer-Based Evaluation and Clinical Management System," paper presented at the Symposium on Microcomputer Applications in Developing Countries, Colombo, Sri Lanka, 1984.

World Bank. "A Methodology for Regional Assessment of Small Scale Hydropower," Energy Dept. Paper no. 14, Washington, D.C., May 1984.

World Development Report. The World Bank, Washington, D.C. 1983.

Index

Africa, 12, 45, 67, 69, 85, 142, 143, 146, 155, 162
Agency for International Development (AID)
Funding for projects, 11, 16, 49, 67, 75, 81, 85, 132
Sponsored or contracted programs, 33, 51, 61, 70, 121, 127,
APMEPU, Nigeria, 54
ASCII, 32, 93
Asian Institute of Technology (AIT), 9
Australian National University, 42

Bangladesh, 12, 73, 94, 95, 102, 18
BASIC
Programming language, 10, 11,
Programs written in, 39, 46, 78-80, 128, 134,
BIOCUT, 127, 151-53
BMASS, 136
BMDP, 102-03
BRANCH, 79-80
Brazil, 60, 156, 162-63
Budgets, 16, 25, 28, 42-43, 50-52, 85,
Bureau of the Census, 45, 70
Burkina Faso, 68

California Institute of Technology, 161
Cameroon, 143
CARINET, 31
CASNET, 83
Centers for Disease Control (CDC), 54, 90, 92
Chile, 81, 162
Colombia, 12, 162
Communications Between microcomputers, 22, 25-26, 29-35, 49, 172, 183
Between microprocessors, 140-41
Policy on, 175
Compendex database, 60
Computer-aided design (CAD), 106, 114, 116, 118, 124
design and drafting (CADD), 114. 116, 161
engineering (CAE), 114, 116
graphics, 112, 114
instruction (CAI), 41
Costa Rica, 133
CP/M operating system, 6, 50, 145, 147
Cuba, 48

Database management software see also individual names, 20, 22-24

Databases, examples of use
    26-27, 32, 45, 59-61, 67,
    90, 94-95, 121, 139, 188
D-B Master, 85
dBase II, III, 90, 145
DENDRAL, 184
Department of Agriculture, 42,
    51
Department of Energy, 106, 112,
    130
DEVELOP database, 27, 61
DIALOG, 32
Digitizer, 141
Dominican Republic, 11, 106
Drinking Water and Sanitation
    Decade, 77

EFAM, 136, 138
Egypt, 3, 12, 45, 106
Electronic Information Exchange
    Service (EIES),30-32
Electronic worksheet, see
    Spreadsheet
Energy audit, 121
ENMAC, 136
Exerpta Medica database, 60

FARMAP, 42
Fiji Islands, 162
Florida, University of, 39, 40
Food and Agriculture
    Organization (FAO), 42
FORTRAN
    Programming language, 11
    Programs written in, 39, 79,
    138, 147
Fuelwood model, 127, 136, 151

Geographical Information
    Systems (GIS), 156,
    160-62
Georgia Institute of Technology,
    12, 134

Harvard Institute for Interna-
    tional Development,43
Hungary, 48
HYDRO-CALC, 130
HYDRO-ECON, 130, 131
HYFEAS, 128

IEEE-488 interface, 144
India, 45, 48, 76, 78, 80, 145-48
Indonesia, 10, 45, 54, 55, 68, 78,
    80, 162
Industrial Research and
    Development Institute of
    Central America
    (ICAITI), 121
Interactive Statistical Inquiry
    System (ISIS), 67, 76
Inter-American Development
    Bank (IDB), 12
International Centre for
    Diarrheal Disease
    Research, Bangladesh
    (ICDDR,B), 94
International Council for
    Research in
    Agroforestry (ICRAF),
    42
International Development
    Research Centre (IDRC),
    43
International Science and
    Technology Institute
    (ISTI), 61
INTERNIST, 74
Ivory Coast, 85, 143

Japan, 31, 48, 170, 186
Johns Hopkins University, 60, 61

Kenya, 43, 49-51, 106, 142, 143
Korea, 31, 106, 162

Liberia, 151, 152
LOOP, 78, 80, 92

LOTUS 1-2-3, 12, 66, 75, 76, 79, 122, 134, 138, 143, 145

Maintenance and service
Providing for, 5, 7, 29, 138, 162, 170-71
Problems with, 13, 29, 134, 155
Malaya, University of , 87, 90
Mauritania, 45
Mauritius, 162
Medline database, 60
Mexico, 32, 60, 162
Michigan State University, 28, 42-44
Microprocessors, 6, 46, 48, 145, 160
Modelling, 11-12, 48-50, 72-77, 108, 112, 124, 126, 139-40, 160-61
Examples of, 12, 40-41, 66, 72-77, 95, 135-38, 153, 156, 162
Software for, 72-77, 109, 113, 116, 120-21, 125, 127, 147, 151-54
see also Simulation
Modems, 25, 29, 32, 49, 183
Morocco, 11, 67, 76, 106, 107, 132, 133
MS-DOS operating system, 6, 136, 145, 147
MULBUD, 42
MYCIN, 74

NASA, 35, 161
Nepal, 53, 16, 108, 143
Nigeria, 45, 54, 143
NIPSOM, 102, 103

Oklahoma Climatological
Survey, 43
ONOCIN, 74

PACSAT, 34, 35
PAPER-ECON, 130, 131
Paraguay, 162
Pennsylvania, University of,66, 70
Peradeniya, University of, 62, 71
Peru, 33, 106, 162
PFS:file, 102
Philippines, 31, 78-80, 106, 144, 162
Pittsburgh, University of, 74
PLATO, 40
Policy
Microcomputer support for - decisions, 3, 16, 18-20, 25, 61, 74, 108, 165-82
Programs that consider - concerns, 73, 75, 92, 94, 127, 135-36
POPLINE database, 60, 61
Population Studies Center, 66, 67, 70
Portugal, 51, 106
Power supply, 4, 8, 13, 29, 50, 64, 172
PROCALFER, 17, 51, 52
Process control, 105, 106, 123, 144
Programmable controllers, 123
Programmable calculator, 13
Projects Management Center, 15
PROSPECTOR, 185

REFLP, 136
RESGEN, 136, 138
RS-232C interface, 144
Research Triangle Institute
(RTI), 68, 73-75, 156, 158, 162
Rapid II, 70, 75, 76

SAS, 40, 70
Sahel, 60, 61, 67
Senegal, 142, 143

Service see Maintenance and
 service
SHDS, 85, 86
Sierra Leone, 45, 85
Simulation, 11, 25, 72-74, 160,
 184
 Examples of, 40, 48, 60, 71,
 84, 99, 136, 139
 Software for, 116-17, 120,
 146, 148-51
 see also Modelling
Software development, need for
 local, 9, 13, 14
Spreadsheets, 12, 22, 24-26, 75,
 76, 122, 134, 145
SPSS, 70
Sri Lanka, 62, 71, 78, 80, 121,
 135, 136, 138, 147, 167,
 177-82
SRS, 94-97
St. Anthony Falls Hydraulic
 Laboratory, 128
Statistical analysis
 Examples of, 28, 44-47, 54, 59,
 67-70, 76, 87, 134
 Software for, 40, 70-71, 102,
 134
Sudan, 106, 133-35
Sukamandi Food Crops Institute,
 55
Supercalc, 79, 134
Swaziland, 64, 162

Taiwan, 31, 55
Tanzania, 73, 108
Technology transfer, 4, 8, 10,
 121, 168, 169
Telenet, 30-32
Telex, 30, 32, 33

TELPLAN, 40
Terrania, 12, 143
Texas A&M University, 28
Thailand, 10, 31, 48, 62, 68,
 78-80, 92, 97
Training
 Examples of, 8-13, 28, 68, 70,
 78-80, 85-87, 102, 142-43,
 Need for, 7, 15, 20, 29, 67,
 108, 138, 155-56, 162, 186
Tunisia, 10, 43, 45, 68, 73, 106,
 122, 158, 160, 162
TWOZONE, 147-49, 151
Tymnet, 32

U.S. Geological Service (USGS),
 161

Value-added networks, 30-32
 see also Telenet, Tymnet,
 CARINET
Versaform, 123
Video disc, 21, 22, 188
VisiCalc, 50, 122, 134
Volkswriter, 134
Volunteers in Technical
 Assistance (VITA), 12, 30,
 31, 34, 61

Water and Sanitation for Health
 (WASH), 61
Word processing, 6, 22, 25, 26,
 50, 134
Wordstar, 134
World Bank, 12, 15, 73, 78-80,
 106, 128, 132, 162

Zimbabwe, 162